The Principles of Modern Biology

The Working Plant

D. A. COULT, M.Sc.

School of Agriculture, University of Nottingham

LONGM.

LONGMAN GROUP LIMITED
London

Associated companies, branches and representatives
throughout the world

© *Longman Group Ltd* 1973

First published 1973
Fourth impression 1979

ISBN 0 582 32319 3

Printed in Hong Kong by
Sing Cheong Printing Co Ltd

Acknowledgements

I should like to thank the many people who have contributed in their several ways to the completion of this book.

I gratefully acknowledge help from my colleague Dr. L. G. Briarty who contributed a short section (and Fig. 8 · 1) on transfer cells, and whose photographs in Chapter 4 are indicated by the initials L.G.B. in the legends.

A great deal of argument and discussion has led to changes here and there in the presentation, but special tribute is due to Harry Grenville (a collaborator in this series) and to my former colleague John Hillman, now in the University of Glasgow; to these and others I am most grateful, though I must nevertheless accept full responsibility for the opinions now expressed. I should be grateful to be notified of the errors which remain, whether of fact or interpretation, so that future versions may be improved.

Much of the burden of the typing has fallen on my sister Jean Coult, whilst Cynthia Clarke has succeeded in making elegant bricks out of the straw of my rough sketches; my warm thanks are due to both of them.

We are grateful to those listed below for permission to reproduce or redraw illustrations for the following figures:

Fig. 3 · 4a Esau, K. (1936). Vessel development in celery (after plate 4). Hilgardia 10: 479, Univ. of California Press.

Fig. 3 · 6c, d Esau, K. (1953). *Plant Anatomy* (after Fig. 12 · 11) John Wiley and Sons, Inc.

Fig. 5 · 6 Curtis, O. F. and Clark, D. G. (1950). *Introduction to Plant Physiology* (after Fig. 7 · 5). McGraw-Hill Book Co.

Fig. 5 · 7 Myers, J. and Burr, G. O. (1940). *Journal Gen. Phys.*, **24**: 45–67.

Fig. 5 · 8 Gaastra, P. (1959). Photosynthesis of crop plants as influenced by light, CO_2, temperature and stomatal diffusion resistance. *Med. Land. Hoogesch. Wageningen*, **59**: 1.

Fig. 5 · 9 Curtis, O. F. and Clark, D. G. (1950). *Introduction to Plant Physiology* (Fig. 2 · 10). McGraw-Hill Book Co.

Fig. 6 · 1d Mühlethaler, K. (1950). Electron microscopy of developing walls (Fig. 8a). *Biochim. Biophys. Acta*, **5**: 1–9.

Fig. 7 · 7 Sutcliffe, J. F. (1962). *Mineral salts absorption in plants* (after Fig. 14a). Pergamon Press.

Acknowledgements

Fig. 7·8   Beevers, H. (1961). *Respiratory Metabolism in Plants*, (Fig. 20), Row, Peterson and Co.

Fig. 8·2a, b   Esau, K. (1953). *Plant Anatomy* (after Fig. 12·5).

Fig. 8·4   Esau, K. (1969). The Phloem (plate 1) *Handbuch der Pflanzenanatomie*, 5, pat. 2, ed. Zimmerman, W. Borntraeger.

Fig. 8·5   Curtis, O. F. and Herty, S. D. (1936). *Am. Journ. Bot.*, 23: 528–532.

Fig. 8·6   Biddulph, O. (1959). Translocation of Inorganic Solutes. In *Plant Physiology* (ed. Steward, F. C.) vol. 2, Chap. 5. Academic Press.

Fig. 9·2   Burges, A. (1958) *Microorganisms in the soil* (Fig. 4). Hutchinson.

Fig. 11·1   Skoog, F. and Miller, C. O. (1957). Chemical Regulation of Growth and Organ formation in Plant Tissues (plate 3). *Soc. Exp. Biol. Symposia* 11.

Fig. 12·1   Clowes, F. A. L. (1961). *Apical Meristems* (plate 25). Blackwell Scientific Publications.

Fig. 12·6   van der Ween, R. and Meyer, G. (1962). *Light and Plant Growth* (after Figs. 30, 31). Phillips Technical Library.

Fig. 13·1e, h   Audus, L. (1965). Fig. 3 in *Plant Growth Substances*. Leonard Hill.

Fig. 13·2   Kefford, N. P. (1955). *J. Exp. Bot.*, 6, (17): 248. Fig. 1 O.U.P.

Fig. 15·1   Leopold, A. C. (1964). *Plant Growth and Development* (after Fig. 20·4). McGraw-Hill Book Co.

Fig. 15·2   Salisbury, F. B. (1963). *The Flowering Process* (after Fig. 5·8). Pergamon Press.

Fig. 15·6   Nitsch, J. (1965). Physiology of Flower and Fruit Development (Fig. 18). *Encyclopaedia of Plant Physiology*, (ed. Ruhland, W.), 15 (1): 1537. Springer Verlag.

Contents

Contents

Contents

Foreword to the Series

Within the last two or three decades Biology has made a most impressive growth spurt, and biologists have moved forward to a new understanding of some of the major problems of life. The tools that have made these steps possible have largely been made available by contemporary advances in the disciplines of physics, chemistry, and mathematics, and their application in the precise skills of engineering. It has been possible to establish a closer unity between the several biological sub-disciplines, and the integration achieved has enabled substantial progress to be made.

A change like this in the climate of understanding is bound to be reflected in all corners of the biological field, and not least (though often last) in its more general literature. This present series of monographs arose out of more than two years of discussions concerned with the 1966 revision of the Northern Universities Joint Matriculation Board syllabus in Biology at Advanced level. It seemed right that out of this conversation should emerge an attempt to provide (in its modern context) working material for the new syllabus.

The series is tied much more to the spirit of this syllabus than to its text; in this sense especially an attempt has been made throughout to emphasise biological principles, so that these books will be suitable for those who are in their preliminary year at universities and colleges.

D. A. COULT

To my teachers T.G.H. and E.J.S.

Introduction

This book deals with some aspects of the structure and functions of the vascular seed-bearing plants (*Spermatophyta*) that dominate much of the earth's surface. The Spermatophyta include the *Angiospermae* (or flowering plants) and the *Gymnospermae* (of which the best known representatives are the conifers). We shall be mainly concerned with the flowering plants.

From an evolutionary point of view it is thought that higher plant life has gradually emerged from an aquatic or subaerial to a drier aerial environment, and considerable adaptation has taken place. Certain features of adaptation have assumed major importance, and some of the earliest problems to be solved must have included:

i. The development of a more efficient photosynthetic system.
ii. The control of water loss to the atmosphere from above ground organs.
iii. The retention of contact with soil water (and nutrients) by the development of a root system.
iv. The development of an adequate transport system to serve and connect root and shoot and other parts of the plant where metabolic activities are taking place.

By virtue of improved photosynthetic efficiency, a more extensive trapping of solar energy has made greater growth and development possible, thus improving the prospects of survival. The system that has arisen as leaves have evolved has allowed for a better supply of carbon dioxide to protected photosynthetic cells, but it has also greatly increased the risk of desiccation through the outward movement of water vapour over the same route. The reverse also holds, for if a system of control of water loss is to be effective it must not too greatly hold up the process of carbon dioxide assimilation.

More capital means more potential growth and increase in size. The seed plants have achieved this by the secondary development of supporting and conducting tissues, making it possible to maintain a much larger canopy of leaves. Some of the extra capital has been diverted into seed production and other devices for the spread of the plant population.

All of these features are inevitably interlinked, and it is necessary to be somewhat arbitrary as to the order in which they are discussed.

Introduction

So we start in Part 1 by looking at the innate properties and problems of plant cells in general, reminding ourselves briefly about the metabolic processes whereby they obtain and deploy energy for the promotion of synthesis, growth and differentiation. Then we go on to look in summary at the form and structure of typical higher plants; this includes a broad but not detailed survey of the distribution and integration of tissues within the plant body, together with some of the main variations encountered in the structure of leaf, stem and root.

In Part 2 which deals with the functional processes and problems of the plant, it has been thought important to underline the principle that structure and function are always intimately related. For example, detailed discussion of leaf anatomy has been deliberately transferred to the chapters on photosynthesis and transpiration, whereas detailed consideration of phloem tissues should obviously be linked up with the discussion of translocation. Similarly, aspects of root anatomy have been considered in relation to water and ion uptake, just as discussion of xylem structure has been related to water transport.

Part 3 reviews some of the processes that result in growth, and it describes differentiation and growth in relation to external and internal factors.

Parts 4 and 5 consider the biology and physiology of reproduction, dispersal and survival; it is in this area that we have to look for the major intervention of the forces of selection, and for some of the adaptive features of structure and metabolism which have helped the flowering plants to survive, in the face of continuing environmental change, through to the present day.

PART ONE

Cells, tissues and organs

1 Features of plant cells and tissues

1 · 1 *Plant and animal cells and tissues*

Cell biologists are agreed as to the similarity in fine structure between plant and animal cells. Survey pictures made with the electron microscope show very similar organisation of nucleus, endoplasmic reticulum and ribosomes, mitochondria, Golgi bodies and other organelles in both plant and animal cells. In addition to the limited discussion of subsection 1 · 2 : 2, a fuller account of these features may be found in *Molecules and Cells*, previously published in this series.

Essential differences, however, may be seen in the outer bounding layers of plant and animal cells. In higher plant cells the outer membrane or *plasmalemma* lays down a relatively rigid *cellulose wall* externally to itself. After mitosis in animal cells, nuclear separation is followed by furrowing (invagination of the equatorial surface) and cleavage into two daughter cells. In plant cells, a *phragmoplast* or cell plate is established between the two daughter nuclei, in and upon which are deposited the materials of the middle lamella; subsequently cellulose microfibrils are added from either side. Furthermore most plant cells develop an extensive vacuolar system, which in conjunction with the cellulose envelope greatly assists in giving support to the plant body by the development of cell turgor (see 1 · 2 : 3).

Plant cells are also unique in developing plastids, chief amongst which are chloroplasts; these contain the green pigment chlorophyll upon which the whole living world depends (see 2 · 2 and Chapter 5).

Animal cells need also to maintain turgor, but whilst the turgidity of animal cells contributes to the 'tone' and adequate functioning of the tissues concerned (e.g. muscle), the higher animal depends for support and movement on a hinged bony skeleton.

The stationary plant makes use of structural woody tissues which lend it mechanical support, but it is the combination of turgid

3

tissues and mechanical tissues that is so important in keeping leaves and flowers properly displayed. Whilst it is true that animal cells also secrete material outside their plasmalemma, it is uncommon except in skeletal tissues for this material to assume the structural importance of the cellulose wall in its various modifications.

Plant tissues are generally better ventilated than animal tissues; they possess a better permeating airspace (or inter-cellular space) system. Even such an apparently solid storage tissue as that of a potato tuber has intercellular spaces that interpenetrate its cells to the extent of 1% by volume, and this seems to allow oxygen to reach the innermost cells at a concentration that is adequate for aerobic respiration. By contrast, some aquatic flowering plants may have as much as 70% of their volume occupied by airspaces; it is also interesting to note that most roots have an extensive airspace system that runs well forward close to the meristematic tip (see Fig. 4 · 10d). Gaseous diffusion is at least 10000 times as fast as diffusion in aqueous solutions, and the advantages of possessing any sort of interpenetrating system of airspaces are obvious. It is worth underlining a comparison in these terms between the tissues of higher plants and animals. Gaseous diffusion aided by mass flow brings oxygen to the alveolar membranes of the mammalian lung; thereafter oxygen movement (and carbon dioxide return) is by way of a liquid phase. The disadvantages in the animal of close-packed tissues without airspaces are overcome by the device of a penetrating circulatory system, combined with the oxygen-carrying properties of haemoglobin. Both plant and animal can cope with their normal supply situation; however, in the plant an oxygen debt may develop when the diffusion pathway becomes so long that most of the oxygen is consumed en route, whereas in the more mobile animal an oxygen debt is nearly always the result of an overall increase in oxygen consumption, such as occurs after extra exertion.

1 · 2 *The boundaries of plant cells and their significance*

1 · 2 : 1 The cellulose wall

All cells need to be considered in relation to their environment, and in a tissue this is most often provided by neighbouring cells. Movement of water and solutes (including dissolved respiratory gases) to and from a cell involves the process of aqueous diffusion. The rate of diffusion of substances in water generally decreases with lowered temperature, even in the case of such a substance as oxygen whose solubility in water increases with lowered temperature. A good approximation is that diffusion rates decrease by about 1 % for every °C decrease in temperature. Aqueous diffusion is important to a plant cell in terms of (*a*) its water relations, (*b*) the passage towards it of solutes and (*c*) the passage away from it of manufactured or synthesised products and waste substances.

The cellulose microfibrils of the cell wall are embedded in a jelly-like matrix of hemicellulose and pectins, and the wall thus constituted can act as a rate-limiting barrier to the diffusion of larger molecules such as sucrose. For example, a leaf cell of the water plant *Elodea* takes longer to come into equilibrium with a plasmolysing solution of sucrose than with an osmotically equivalent solution of KCl. However, for the most part the cell wall is not an important barrier to the passage of solutes, and we have to look elsewhere for the occurrence of a means of solute control. The important changes that may take place in the cellulose wall include thickening, by the incorporation of new cellulose microfibrils into the wall, and subsequent *lignification, suberisation* and *cutinisation*.

LIGNIFICATION involves changes in the matrix materials which surround the framework of primary cellulose micelles in a cell wall, and polymerised aromatic molecules are laid down. This process of lignification follows after the laying down of primary cellulose, and results in an intimate association of lignin and cellulose called ligno-cellulose. The physical properties of lignified cellulose differ from those of unlignified cellulose in respect of reduced elasticity (i.e. increased rigidity and therefore reduced flexibility). This is why unlignified collenchyma (p. 21) is so useful as a supporting tissue in developing young stems and leaves, whereas later on, tissues in the mature stem need to be more rigid. However, it must be stressed that a lignified wall is not so much less permeable to water than a comparable unlignified wall.

SUBERISATION and CUTINISATION are similar processes and involve the impregnation of a primary or secondary cellulose wall with polymerised lipidic substances, associated more or less with waxes. Cutin and suberin are semi-hydrophilic substances, and can

5

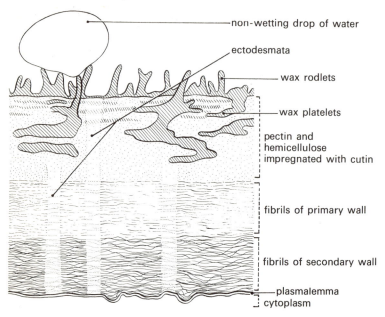

non-wetting drop of water

ectodesmata

wax rodlets

wax platelets

pectin and
hemicellulose
impregnated with cutin

fibrils of primary wall

fibrils of secondary wall

plasmalemma
cytoplasm

Fig 1·1 An interpretation of the structure of a cutinised cell wall with cuticle. Water can pass slowly across the cutinised part of the wall, but movement is restricted by the presence of wax. The pores marked ectodesmata offer little impedance to water movement in either direction, but they do not break the continuity of the plasmalemma on the inside, nor of the waxy layers on the outside.

thus establish contact with hydrophilic cellulose on the one hand and with waxy materials (hydrophobic) on the other. The semi-hydrophilic nature of cutin itself makes it possible for water to move through it, but water movement is very much retarded by the presence of organised wax layers (Fig. 1 · 1); however, as we shall see (in section 5 · 2 : 3), cuticular transpiration is rarely wholly cut down to zero.

1 · 2 : 2 Cell membranes

Most cells behave as if they were surrounded by an external membrane whose properties may vary according to circumstances. The cellulose cell wall is laid down externally to the membrane and as part of the membrane's activity. The membrane can be detected with the electron microscope; it is known as the *plasmalemma* and is described as a lipo-protein membrane. It is differentially permeable to solutes, and can be shown to vary in its permeability to water. For example the epidermal cells of young bulb-scales of onion allow the passage of water more readily in summer than in winter; again, some parts of a cell may often seem to be more permeable to water than the remainder.

So far as the passage of solute molecules is concerned, it can be said that substances with an affinity for fats, e.g. fat solvents, pass easily across the membrane; polar molecules like sucrose, which has eight hydroxyl groups (OH) per molecule, penetrate with more difficulty. Charged particles, anions and cations, have to be 'ferried' across (probably in combination with what is known as a carrier molecule); this can only be done with the expenditure of energy, and it is found that a salt-starved tissue shows an increase in respiration when bathed in an electrolyte solution.

The 'health' of a membrane, and its continued capacity to discriminate between various ionic species thus seems to depend on the energy made available for its maintenance during respiration, and more must be said on this subject when the functions of the root are being described.

The membrane systems within the cell are thought to serve a variety of functions according to their location. The movement of materials into the vacuole (containing cell sap) is controlled by the *tonoplast*, a membrane that bounds the vacuole. The accumulation of solutes in the vacuole may be important as a means of storing needed materials, such as K^+ ions or sucrose; it may also be a useful way of setting aside some of the waste products of metabolism.

Similar membranes isolate the so-called *organelles* of the cell from the remaining clear cytoplasm. These include: the *mitochondria*, in which much of the energy required by the cell is made available for its many metabolic processes; the *plastids* (and especially the green photosynthetic chloroplasts); and the nucleus itself, the repository of stored information for the management of the activities of the cell. Where such an organelle is doing a specific job or jobs, it is clearly valuable to isolate it from the remaining parts of the cell. The composite diagram of a young leaf cell of the pondweed *Elodea* serves to bring together and display some of these features (Fig. 1 · 2). Brief reference must also be made to the *endoplasmic reticulum*, a membrane-bounded canal or cavity system which may help to bring the external environment into closer contact with the inner parts of the cell. The membranes may sometimes be studded on the outside with *ribosomes* where stored information from the nucleus is 'translated' into protein manufacture;

7

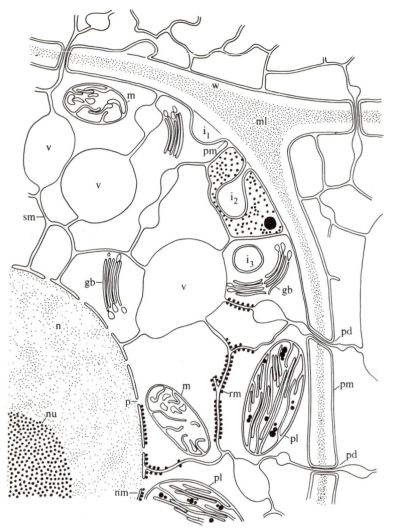

Fig 1·2 Diagram showing the characteristic structure of a meristematic cell of *Elodea sp.*, the pond weed. Note the nucleus (n) with its nucleolus (nu) and double nuclear membrane (nm) penetrated by pores (p); the cavity of the nuclear membrane is continuous with the endoplasmic reticulum. The smooth membrane (s) of the endoplasmic reticulum locally balloons out into vacuoles (v).

The plasmalemma (pm) bounds the outermost part of the protoplast; it is a typical 'unit' membrane. Invaginations of pinocytotic origin are seen in a series (i_1, i_2, i_3). The primary cell wall (w), adjacent to the plasmalemma, shows a middle lamella (ml); it is penetrated by plasmodesmata (pd) which allow contact between adjacent cells via their endoplasmic reticular systems. Other organelles include plastids (pl), mitochondria (m), ribosomes on the rough membrane (rm) of the endoplasmic reticulum, and Golgi bodies (gb).

here are made some of the working promoters of metabolism, the enzymes. Note that the ribosomes may sometimes occur lying freely in the clear cytoplasm, they are not membrane-bound even though they are often associated with membranes.

These and related topics are dealt with more fully in the author's *Molecules and Cells* in the series.

1 · 2 : 3 Osmosis and turgor

Another outcome of the restraining presence of the plasmalemma and tonoplast is that the cell acts as an osmotic unit. The unit can be thought of in this respect as a vacuole bounded by membranes

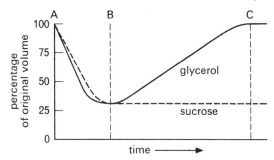

Fig 1·3 Graph showing the change in volume of *Spirogyra sp.* protoplasts plasmolysed in isosmotic solutions of glycerol and sucrose respectively. Time AB could be used as an estimate of the rate of water loss during plasmolysis; time AC, including plasmolysis and full recovery, may be used to estimate the rate of entry of glycerol, and to compare it with that of other substances. Sucrose does not appear to enter the cell at all.

and cytoplasm, together acting in such a way that whilst water enters and leaves quite readily, most solutes do so very much more slowly; in some cases this entry is so slow as to be hardly perceptible. For example, if leaf cells of *Elodea*, or if threads of *Spirogyra* are plasmolysed in isosmotic solutions (i.e. solutions which have the same osmotic potential) of (*a*) glycerol and (*b*) sucrose, protoplast length or volume may then be measured and plotted against a time-base (Fig. 1 · 3). It will be seen that after fairly fast contraction of the cell, there is a slow recovery towards the original volume, as glycerol molecules slowly penetrate the membrane and help in the osmotic readjustment of cell volume. Not so in the case of the sucrose solution, where the final equilibrium plasmolysis volume stays low and shows little signs of recovery over quite a considerable period.

9

Water molecules move from regions where the H_2O concentration is high to regions where H_2O concentration is low; so also do solute molecules in an aqueous solution. If a restraint is placed on the movement of the solute molecules in a vacuole by the presence of an osmotic membrane barrier, whilst water molecules are able to move across the barrier quite fast, the solute molecules can at best move across slowly. If the concentration of solutes within the vacuole is higher than the concentration of solutes outside, water will move inwards, i.e. from a region of higher water and low solute concentration to a region of lower water concentration because of higher solute concentration.

The concentration of water in a given solution can be expressed as its *mole fraction*, this being given by the ratio:

$$\frac{n_{H_2O}}{n_{solute} + n_{H_2O}}$$

where n represents the number of moles present of each species. Thus a 5% (w/w) solution of sucrose (i.e. 5 g of sucrose dissolved in 95 g of water) contains 5/342 moles of sucrose (since the molecular weight of sucrose is 342) and 95/18 moles of water. The mole fraction of water present is:

$$\frac{95/18}{5/342 + 95/18} \quad \text{that is } 0.998$$

In a 5% w/w solution of an electrolyte such as potassium nitrate, which dissociates fully into its constituent ions

$$KNO_3 \rightarrow K^+ + NO_3^-$$

there are

$$5/39.1 \, (= 0.128) \text{ moles of } K^+$$
$$5/62.0 \, (= 0.081) \text{ moles of } NO_3$$

and

$$95/18 \, (= 5.28) \text{ moles of water}$$

The mole fraction of water here is

$$\frac{n_{H_2O}}{n_{H_2O} + n_{K^+} + n_{NO_3}} = \frac{5.28}{5.49} = 0.961$$

Thus the effective concentration of water in a 5% solution of sucrose is greater than that of the water in a 5% KNO_3 solution, and if two such solutions were separated by a semi-permeable membrane, permeable only to water but not to other solute molecules or ions, water would flow from the sucrose solution to the nitrate solution across the membrane until equilibrium was reached.

We have assumed of course that solvent and solute particles are here acting independently of one another. In fact a correction must be made because a molecule such as sucrose may attract water molecules to itself in such a way as to render some of them unavailable for independent action; we say then that their activity (or effective concentration) has decreased. It is similarly a fact that the highly charged ions also influence to a greater or lesser degree the ' halo ' of water around them and thus bring about departures from the normal, unless we are dealing with solutions at great dilution.

Osmotic flow is thus determined by the respective concentrations of water present on each side of an osmotic membrane, and this in its turn is dependent on the relative numbers of solute particles (molecules or ions) and upon their molecular (or ionic) weight. For a given percentage concentration sucrose (molecular wt. 342) is less effective osmotically than glucose (m. wt. 180), and glucose is less effective than KNO_3, giving potassium (ionic wt. 39·1) and nitrate ions (i. wt. 62·0). In general it is a convenient rule to remember that one mole (i.e. the g mol. wt.) of a given solute made up to a litre of solution should at N.T.P. theoretically have an osmotic potential of 22·4 atmospheres (2269 kN/m^2); but it will deviate from this figure to an extent which will depend primarily on the interaction between molecules or ions that we have mentioned above. For example, at 20°C a molar solution of sucrose (342 g made up to a volume of one litre with water) has an osmotic potential of nearly 35 atmospheres and the deviation is considerable in stronger solutions. However, it is convenient for calculation to use this analogy with the gas laws, though it must be stressed that, whilst thinking in terms of solute concentration, we must not forget that osmotic water movement is achieved always as a result of flow down a concentration gradient of water.

Thus we can say that $\pi V = nRT\phi$ where π is the osmotic pressure of n moles of solute occupying a volume V, R is the gas constant and ϕ is the osmotic coefficient, which corrects for deviation from the theoretical value and has to be determined experimentally for given solute concentrations. It will be seen that n/V is the molarity of the solution (the number of moles per litre) and since also:

$$n = \frac{\text{no. of g of solute present}}{\text{mol. wt in g of solute}}$$

π must vary inversely with the molecular weight, and directly with the molar concentration of the solute; it also varies directly with the absolute temperature, and will increase as the temperature rises.

In considering the plant as a working unit, we should be interested (from the osmotic point of view) in two main features. The first of these is that water can be moved from one cell to another, individual cells adjusting their water content to that of the neighbours in accordance with the osmotic equilibrium established between them. Similarly they adjust themselves to an external solution which bathes the plasmalemma. This external solution may just be part of the imbibed cellulose wall system external to each protoplast; it may however extend into and be continuous with the soil solution (see under root and soil).

The other feature concerns the *turgor* of the cell. The osmotic contents of a cell vacuole may in fact be at such a concentration that water flows in from the exterior and the vacuolar volume then increases to the point at which the cell wall becomes elastically taut, and resists the outward pressure of the vacuolar system with a balancing inwardly directed *wall pressure*. Cells of tissues which are fully osmotically imbibed are said to be turgid; they provide an extra means of support for tissues, and this implements the mechanical action of the vascular elements present. The well known experiment using split dandelion scape or similar material demonstrates that when tissue compression due to turgor is released, there is an initial rapid change in curvature which increases more slowly as water is taken up by the inner parenchymatous pith cells, and the portion of scape curves outwards, with the epidermal and vascular tissues on the inside of the curve. When the scape is placed in a solution that removes water from the pith cells, its curvature is reduced and may even be reversed.

The German physiologist Janse pointed out that if the young turgid and erect shoots of *Solidago* (Aaron's Rod) are tapped smartly with the finger or a pencil, water is caused to escape from the turgid cells into the intercellular spaces around them, and the shoots droop as if they had wilted. They quickly recover if the water supply is adequate. The importance of turgidity in terms of the mechanical support of the overground shoot needs to be emphasised.

2 The energy relations of plant cells

The foregoing resumé of the characteristics of living plant cells would be unbalanced without some general reference to their metabolic activity; basic to any understanding of metabolic processes is the concept of energy and its utilisation during the normal functioning of cells. Although the standpoint of this book is physiological rather than biochemical, readers should nevertheless be reminded that every physiological process is ultimately dependent upon energy-yielding molecular reactions. The cells of the working plant are of course wholly dependent upon solar energy for all that they accomplish in growth and synthesis. Radiant energy is converted into chemical bond energy, which may subsequently be harnessed to do various kinds of work, e.g. chemical, osmotic, electrical etc.

Only the shortest of outlines may be given in this preliminary review of the molecular processes which occur in plant cells, and result in the availability of energy; their detail must be studied elsewhere.

2 · 1 *Energy transfer*

The living cell is not just a hotch-potch of molecules interacting at random. Probably most of the reactions that take place are localised at specific enzyme sites in different regions of the cell, often in specific organelles. Many of them occur at membrane interfaces, for example in the plasmalemma and in parts of the endoplasmic reticulum, or in mitochondria and chloroplasts. There is a high degree of orderliness in the cell, and this makes it all the more important that, wherever energy can be made available, there shall be some linking means of conveying it to sites of synthesis and reaction elsewhere, for example from mitochondria to membrane-attached ribosomes.

Adenosine triphosphate (ATP) is typical of a group of substances that are important in promoting biochemical reactions and the transfer of energy within the cell. Its structure can most simply be expressed in terms of the units from which it is made:

$$\underbrace{\text{Adenine} - \text{ribose} - \circledP \sim \circledP \sim \circledP}_{\text{adenosine}}$$

Adenine is a purine base, and it is linked with the 5-carbon sugar ribose to give the nucleoside adenosine. The progressive linking-on of phosphate [$\circledP$] groups to adenosine gives rise to the nucleotides adenosine monophosphate (AMP), adenosine diphosphate (ADP) and adenosine triphosphate (ATP). More energy is required to establish a pyrophosphate bond [$\circledP \sim \circledP$] than the single phosphate bond [$-\circledP$] and therefore the conversion of AMP to ADP, and of ADP to ATP is most likely to occur where any reaction is taking place that has a high enough energy yield. ATP-generating reactions are linked especially to the oxidative processes of respiration based on the mitochondria (*substrate phosphorylation*), but ATP is also generated in the chloroplasts during photosynthesis (*photophosphorylation*).

The pyrophosphate bonds have been referred to loosely as ' high-energy ' bonds. In fact, each newly added pyrophosphate bond increases the internal energy of the whole nucleotide molecule, and when a phosphate group is transferred (or shed) from the nucleotide, some of this energy can be made available to do work (chemical or otherwise). Thus ATP acts as a source of energy for a multitude of chemical syntheses; it is also known to be implicated in ion transfer, which involves the performance of electrical work. Some of the energy may just be dissipated as heat (kinetic energy), as in the classical experiment in which the temperature rises when seeds are germinated in a thermos flask.

It is probable that ATP promotes energy-requiring reactions by becoming joined on to the surface of an enzyme (a *kinase*) in close proximity to the reactant molecules. Then a redistribution of the internal energy of the complex takes place, and enough energy is made available to bring about reaction between the two molecules, after which they can be released in their new form from the surface of the kinase. A simple synthesis might be represented as follows:

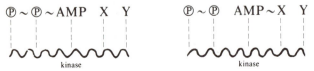

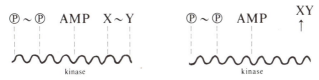

Note here that ATP has been shown as splitting to AMP + pyrophosphate (i.e. ℗ ∼ ℗) though it might equally well split to ADP + ℗. A typical phosphorylation may be shown as follows:

$$\text{Glycerol} + \text{ATP} \rightarrow \text{glycerol phosphate} + \text{ADP}$$

Pyrophosphorylation may take place in a similar way. ATP also serves to phosphorylate other nucleotide phosphates, and phosphate transfer seems to take place readily between most nucleotides. The following three reactions can serve as an example, involving first the phosphorylation of the nucleotide uridine diphosphate (UDP) (derived from the pyrimidine base uracil) followed by the synthesis of the disaccharide sucrose:

(1) UDP + ATP → UTP + ADP
(2) UTP + Glucose 1-phosphate → UDP-glucose + ℗ ∼ ℗
(3) Fructose 6-phosphate + UDP-glucose → sucrose phosphate + UDP

The prior formation of the sugar-phosphate esters (viz. glucose 1-phosphate and fructose 6-phosphate) also requires ATP, and presumably makes it easier for these molecules to become attached to the appropriate enzyme surface.

Enough has been said to indicate that ATP plays a vital part in the mediation of energy, as well as in what might be called 'oiling the metabolic wheels' through phosphorylation; it is justly called the universal energy carrier.

2 · 2 *Photosynthesis*

The most significant metabolic process in the plant is photosynthesis. Chapter 5 is about the 'whole plant' view of photosynthesis, discussing the interacting factors conditioning the biophysical and biochemical reactions that occur during photosynthesis in green cells. Photosynthesis is a process in which light energy is trapped and converted into chemical bond energy. Wherever light impinges upon chloroplasts in turgid cells, CO_2 and energy

15

are incorporated into hexose molecules by means of a complex enzyme system. The hexose molecules hold energy in their chemical bonds, and some of this energy can later be released during respiration and used for many of the working purposes of the plant.

The action of light on chloroplasts results in the production of reducing power, in the form of a reduced co-enzyme molecule. (A co-enzyme is a molecule that is closely associated with an enzyme molecule, and which participates in the reaction which is being catalysed.) This particular co-enzyme is known as NADP (which is short for nicotinamide adenine dinucleotide phosphate; whilst there seems to be little good reason at this stage for memorising detail of this kind, nevertheless it *is* important to think of NADP as a molecule that is capable of accepting or donating hydrogen), and it becomes reduced to $NADPH_2$ by accepting hydrogen. The hydrogen comes from the photolysis of water (i.e. the splitting of water by light) and oxygen is released in the process.

$$NADP + H_2O \xrightarrow{\text{light}} NADPH_2 + \tfrac{1}{2}O_2$$

Some ATP also is generated in the illuminated chloroplasts by a process known as photophosphorylation. These two processes together constitute the *light process.*

Both energy (via ATP) and reducing power (using $NADPH_2$) are required in the complex *dark process*, by which CO_2 is incorporated into the system. In effect the CO_2 is reduced according to the following grossly oversimplified statement, and hexose is produced:

$$6CO_2 + 12\, NADPH_2 + \text{energy} \longrightarrow C_6H_{12}O_6 + 12NADP + 6H_2O$$

In the photolysis of water:

$$12NADP + 12H_2O \longrightarrow 12NADPH_2 + 6O_2$$

and it can easily be seen that these two statements can be summed to give the modified classical equation for photosynthesis:

$$6CO_2 + 12H_2O^* \longrightarrow C_6H_{12}O_6 + 6O_2^* + 6H_2O$$

in which the asterisk denotes that all of the oxygen emitted is derived from the photolysis of water, and that none comes from the CO_2 molecule.

Thus energy from the sun has been trapped and stored in the form of chemical bond energy in the glucose molecule; it may be subsequently released during the process of respiration and then harnessed to do work in the cells.

2 · 3 *Respiration*

Just as the processes of energy storage through photosynthesis are dependent on a complex series of enzyme systems, so too the availability of that stored energy for work depends on other enzyme systems. In the cytoplasm of all living cells, perhaps loosely associated with membrane systems such as the endoplasmic reticulum, each hexose molecule may be broken down to two 3-carbon molecules which may then give rise to pyruvic acid (CH_3 . CO . $COOH$). This splitting of the 6-carbon sugar into two 3-carbon molecules is known as glycolysis, and although it can take place in the absence of oxygen, its energy yield is small, as judged by its low production of ATP and thus harnessable energy.

The series of processes involved in glycolysis can be summed as follows:

$$C_6H_{12}O_6 + 2ADP + 2iP \rightarrow 2CH_3 . CO . COOH + 2ATP$$
$$\text{(glucose)} \qquad\qquad \text{(pyruvic acid)}$$

In higher plants, when oxygen is available, the organelles of the cell known as mitochondria come into play, and promote the complete oxidation of pyruvic acid to CO_2 and H_2O, with the production of much more ATP:

$$CH_3 . CO . COOH \rightarrow 3CO_2 + 3H_2O$$
$$15ADP + 15iP \rightarrow 15ATP$$

This latter process is not available in some micro-organisms and they have to make do with the relatively slender output of energy associated with glycolysis.

Higher plants in the absence of oxygen may divert their pyruvic acid molecules into ethanol production. Ethanol (C_2H_5OH) is a moderately volatile substance, and can escape into the surroundings in the same way as CO_2. The biochemical pathways between hexose and pyruvic acid involve the production of reducing power, this time as $NADH_2$, the reduced form of the co-enzyme nicotinamide adenine dinucleotide. In the absence of oxygen this reducing power is linked to the formation of ethanol, and CO_2 is removed at the same time; rather more rarely is lactic acid formed as in animal muscle.

$$CH_3 . CO . COOH + NADH_2 \rightarrow CH_3 . CH_2OH + CO_2 + NAD$$

(pyruvic acid)   (ethanol)

$$CH_3 . CO . COOH + NADH_2 \rightarrow CH_3 . CHOH . COOH + NAD$$

(pyruvic acid)   (lactic acid)

Aerobic respiration, mediated through the enzymes of the mitochondria, produces a lot of energy during the breakdown of hexose sugar molecules; it also wastes more than half of the total available energy, which is dissipated as heat whilst the remainder is used to form ATP.

Thus one major requirement in the working living cell is for hexose or an alternative carbon source of energy. It is transported to the cell either directly from the leaves, or from some storage source such as starch in another cell. With this energy, once it is available, the cell can carry out all the appropriate processes of growth and metabolism for which it is adapted by differentiation.

3 The structure of plants at the cellular level

Before discussing in detail the general functions of higher plants, it is proposed first to give a broad summary account of the cells, tissues and organs that are likely to be encountered. More specialised accounts of individual tissues will be found later at relevant places in the text.

3 · 1 *General morphology*

There are three main categories of vegetative organs that are recognised in the higher plant, viz. roots, leaves and stems. Roots do not normally give rise to any other than similar root structures, though there are well-known exceptions to this rule, e.g. the root suckers of roses, plums, etc. (which are known as *adventitious shoots*). Stems give rise to leaves and buds; the latter are also stem structures and also come to bear leaves. Bud and leaf may be very much modified during flowering, i.e. in the formation of sexual reproductive structures. Adventitious roots arise much more frequently from stems than vice versa, and it need hardly be pointed out that this is of especial horticultural value in the rooting of cuttings (see p. 158). All three categories of organ may become modified in their adaptation to the function of storage, whilst leaves and stems may show considerable adaptation to the problems of water conservation.

At ground level in the higher plant a *transition-zone* connects root and stem, so that what is effectively a centralised rod of supporting and conducting tissue in the root, broadens to give place in the stem to a more peripheralised open-girder type of supporting system, capable, more or less, of bearing the photosynthetic and reproductive organs aloft, and of withstanding the inevitable stresses and strains that are involved in bearing up a system that is much heavier than air, but is constantly exposed to air movement.

19

Thus in general terms the underground root system is concerned with anchorage, and with water and salt uptake (see Chapter 7). The stem is concerned with the support and display of photosynthesising leaves (see 5·2:1) and flowers (Chapter 14). This exposes the plant overground to the dangers of excessive water loss (Chapter 6). The interconnection of root, leaf and stem has made almost obligatory the development of a specialised transport system (Chapter 8).

3 · 2 *Types of tissue*

The adult plant body is made up of tissues which result from the processes of differentiation referred to in Chapter 12. For some purposes adult cells can be conveniently grouped into two categories, *parenchyma* and *prosenchyma*; in each of these categories the cells may be lignified or non-lignified. In practice, distinguishing cells by length is not entirely satisfactory, and whilst adult parenchymatous cells are rarely much more than about five or six times as long as they are broad, and, by contrast prosenchyma cells are nearly always proportionately longer than this, it cannot be denied that these categories may sometimes overlap. Parenchymatous cells are normally ' packing cells ', forming what is sometimes called ground tissue, whilst prosenchymatous cells are more likely to be concerned with the structural problems of support, in which they are aided by the turgor of the neighbouring parenchyma (see 1 · 2 : 3). Similarly, whilst some cells are quick to undergo lignification, others are very late in lignifying, so that categorisation of cells by this character is again only broadly useful, as in the sense in which it is used here.

In addition to tissues in these categories, the conducting tissues, xylem and phloem, are very distinctive both in form and in function

Fig 3·1 Parenchyma.

a. Lignified pith cells from the stem of *Tilia europaea* (lime) showing simple pits [Ph.]

b. Storage parenchyma from the root cortex of *Ranunculus repens* (creeping buttercup). The cells contain starch: some plasmolysis has occurred. [Ph.]

c. Lignified parenchyma in radial longitudinal section (R.L.S.) from a horizontal ray in the wood of *Liriodendron sp.* Note the extensive thickening and simple pits (p). These are living cells and contain starch grains (s) and nuclei (n). [Ph.]

d. Chlorenchyma cell from the leaf of *Helleborus sp.* Note discoid chloroplasts in face and edge view (c), nucleus (n) and the copious air space system (a).

[Ph.] indicates that phase contrast microscopy has been used.

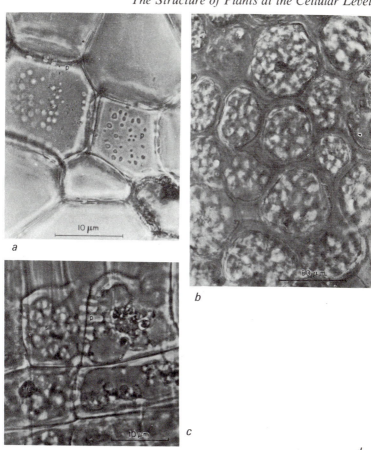

a

b

c

d

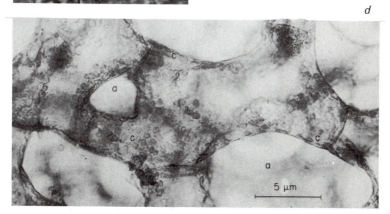

and will be dealt with more fully later in relation to their functions (see 6·1 and 8·1). Nearly all stem and root cells show relative extension in a direction parallel to the longitudinal axis of the plant.

A brief annotated classification of cell types follows and is illustrated in Figs. 3·1–7.

1. PARENCHYMA

If not isodiametric, these cells are rarely much more than five to six times as long as broad. They form ground tissue, e.g. cortex in stem and root, medulla or pith in stem; cortical cells in stem may sometimes be photosynthetic; they are rarely so in the root. Chlorophyll-bearing tissue is known as chlorenchyma. (Fig. 3·1). Parenchyma cells often show relatively little specialisation of form, though considerable differentiation in function (such as starch storage, tannin deposition, photosynthetic activity, support by packing and turgor).

Lignified parenchyma is well typified by the packing cells and ray cells (concerned with transverse transport) in the woody xylem. It may also occur as specialised tough supporting cells (sclereids) in leaves and elsewhere (see Fig. 5·1). Medullary parenchyma often eventually lignifies.

2. PROSENCHYMA

Cells probably (though not always) more than seven to eight times as long as broad, functioning largely in support, and in resisting tension and compression, may at first be collenchymatous (see below) and ultimately become lignified.

Lignified prosenchyma is often referred to as *sclerenchyma* (*scleros* = hard), and is best typified by fibres. Such cells are found in and amongst the woody xylem tissues; they may also be found in the bast or phloem, either protecting each phloem strand with a hard external flattened rod of tissue, or intermixed with phloem cells (as seen in the bark of *Tilia europaea*). (Fig. 4·7d). It is believed by evolutionary anatomists that fibres preceded tracheids and vessels (see below), and since one can encounter fibre-tracheids and other intermediate forms, it must be recognised once again that hard and fast categories are difficult to establish.

The *collenchymatous* state is frequently met with in prosenchymatous cells, though it may also be observed in cortical parenchyma (and even sometimes at the season's end, in the dormant cambial cells of such trees as ash). Collenchymatous cells are characterised by a deposition of translucent hemicellulose materials

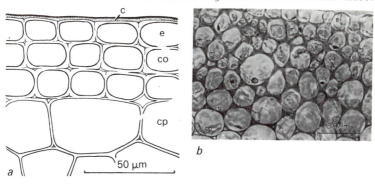

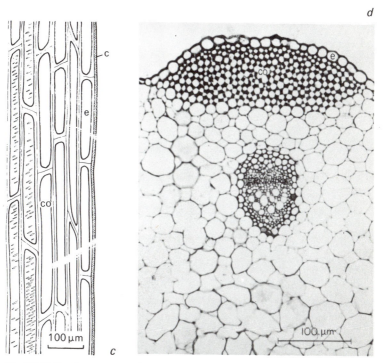

Fig 3·2 Collenchyma.

a. Sub-epidermal collenchyma (co) in a transverse section (T.S.) of *Dahlia sp.*, showing cuticle (c), epidermis (e) and cortical parenchyma (cp).

b. T.S. of epidermis and sub-epidermal collenchyma from the stem of *Ecballium sp.* (squirting cucumber), showing similar structures to 3·2a. [Ph.]

c. Longitudinal section (L.S.) of collenchyma in the stem of *Ecballium sp.* showing simple pits.

d. T.S. of a portion of the petiole of *Rumex sp.* (dock) showing sub-epidermal collenchyma forming a ridge outside the vascular bundle.

23

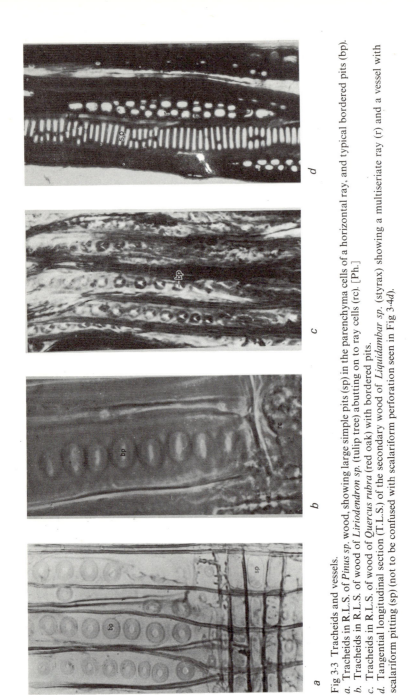

Fig 3·3 Tracheids and vessels.

a. Tracheids in R.L.S. of *Pinus sp.* wood, showing large simple pits (sp) in the parenchyma cells of a horizontal ray, and typical bordered pits (bp).

b. Tracheids in R.L.S. of wood of *Liriodendron sp.* (tulip tree) abutting on to ray cells (rc). [Ph.]

c. Tracheids in R.L.S. of wood of *Quercus rubra* (red oak) with bordered pits.

d. Tangential longitudinal section (T.L.S.) of the secondary wood of *Liquidambar sp.* (styrax) showing a multiseriate ray (r) and a vessel with scalariform pitting (sp) (not to be confused with scalariform perforation seen in Fig 3·4*d*).

The Structure of Plants at the Cellular Level

on to the primary cellulose wall, but in an irregular manner, and largely in bands at the angles of the cell. (Fig. 3 · 2). The hemicelluloses are chemically related to cellulose, but are apparently more labile, i.e. they are more readily deposited and removed. Clearly they may have a storage function. Their presence contributes reinforcement to the wall of an elastic (non-rigid) character. Cells of this kind are often found just beneath the surface tissues of stems and leaves (especially in the main veins of the latter). They frequently become lignified when older.

3. CONDUCTING TISSUES

(a) *Xylem* is the collective term given to the woody (i.e. lignified) conducting tissues, serving mostly for the passage of water and dissolved electrolytes. Xylem may sometimes carry organic materials, particularly in the spring (e.g. sugar maple). The constituent cells of the xylem may vary from species to species, but lignified tracheids, vessels, fibres and xylem parenchyma may be found.

(i) *Tracheids* are best seen in the wood of coniferous stems, though it must not be imagined that they are as clearly recognisable as this in many flowering plants (examples of both are shown in Fig. 3 · 3). The coniferous tracheid is typically chisel-ended, and shows bordered pits especially in its radial walls; it is at first evenly thickened, but may in some cases show the apposition of tertiary strengthening spirals to the secondarily thickened ligno-cellulose wall. Water movement through coniferous tissues is much slowed up because it has to pass from tracheid to tracheid across the thinnest parts of the bordered pits (Fig. 6 · 1).

(ii) *Vessels* are normally only found in angiosperm wood. They are derived from longitudinal files of vessel segments placed end-to-end, in which the early dissolution of each separating end wall gives rise to a perforation, so that there is continuity between the members of the row of vessel segments which go to form a vessel (Fig. 3 · 4a).

The earliest differentiated vessels are easily recognised as part of the protoxylem; they add annular and helical thickenings to their primary walls (Fig. 3 · 5) and are thus capable of keeping pace with the extension growth of the young stem in which they have been laid down. As extension growth slows down, so metaxylem vessels are differentiated; these may show a variety of thickenings to their walls as seen in Figs. 3 · 3d and 3 · 4. Vessel elements that are laid down after stem extension is complete can thus be more strongly thickened and can show more varied patterns of thickening.

25

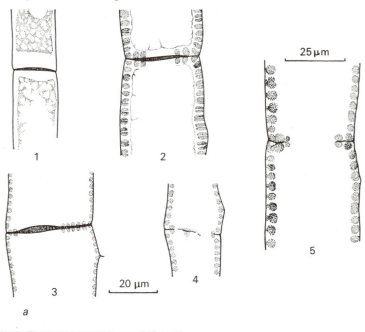

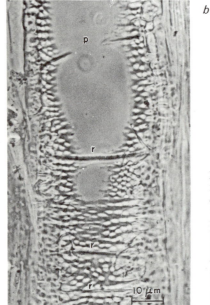

Fig 3·4 Metaxylem and secondary xylem.

a. Stages in the breakdown of the end wall separating adjacent vessel segments in *Apium graveolens* (celery). Note the swelling of the wall prior to perforation. (After Esau.) [Ph.]

b. L.S. of the stem of *Cucurbita sp.* showing a metaxylem vessel. A perforated end wall is seen in section (p), and the remnant rims of other end walls are also present (r).

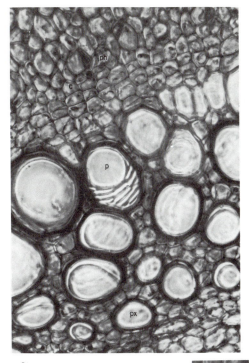

c

d

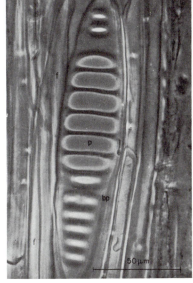

Fig 3·4 continued.

c. T.S. of the stem of *Ecballium sp.* showing a metaxylem vessel with a perforated end wall, the remaining rim bearing elongated bordered pits. Cambium (c), phloem (ph) and protoxylem vessels (px) are also seen. [Ph.]

d. R.L.S. of the secondary wood of *Liriodendron sp.* showing the long, steeply sloping end wall of a vessel, with scalariform perforation (p), bordered pits (bp) and fibres which run on each side of the vessel (f). [Ph.]

27

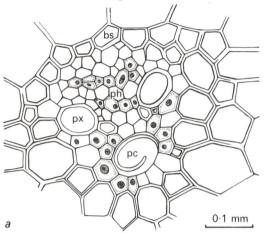

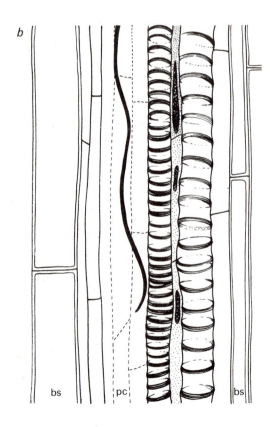

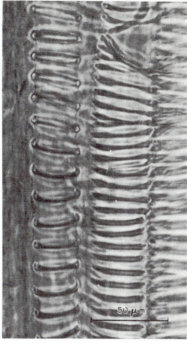

c   d

Fig 3·5 Protoxylem.

a. T.S. of a stem of *Tradescantia sp.*, which is a monocotyledon, showing a protoxylem canal (pc) with the remains of helical thickening from an extended vessel. Other vessels show annular thickening, and phloem (ph) and bundle sheath cells (bs) are also seen.

b. L.S. of the stem of *Tradescantia sp.*

c. L.S. of the stem of *Ecballium sp.* showing a series in the differentiation of protoxylem; protoxylem vessel with helical thickening (px_1) which has been much stretched during stem elongation; a vessel showing much less stretching of the helix (px_2), as growth slows down; and the flattened helix of a late protoxylem vessel (px_3). [Ph.]

d. L.S. of the stem of *Cucurbita sp.* (marrow) showing annular thickening in very early protoxylem. [Ph.]

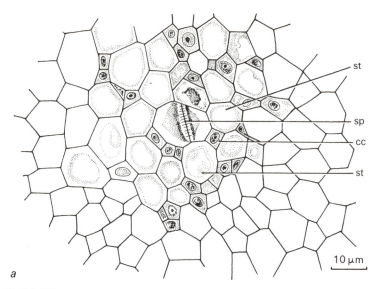

Fig 3·6 Phloem.
a. T.S. of the stem of *Dahlia sp.*, showing phloem with an obliquely sloping sieve plate.
b. L.S. of phloem of *Ecballium sp.*
c. R.L.S. of secondary phloem of *Vitis vinifera* (grapevine) showing sloping sieve plates in face view.
d. T.L.S. of secondary phloem of *V. vinifera*, showing sieve plates in section.
companion cell (cc), sieve tube (st), sieve plate (sp), sieve area (sa), metaxylem vessel (mx), protoxylem vessel (px), callose (cl); (*c.* and *d.* after Esau) (see opposite page).

The xylem of stems and roots is frequently traversed horizontally by *ray* tissue. This is essentially parenchymatous; although lignified it is plentifully provided with communicating pits in its walls. It is a living tissue and frequently acts as a storage tissue for starch and other reserves. (See Figs. 3·1c and 3·3).

(*b*) *Phloem* tissues (Figs. 3·6 and 8·2–4) are concerned mainly but not wholly with the transport of organic molecules such as sugars, amino acids and so on. They are described at greater length in Chapter 8 on translocation, so that all that need be said here is that they include sieve elements and companion cells, phloem parenchyma and phloem fibres. In young tissues and especially in leaf conducting tissues in the finer veins protophloem is found which lacks much of the differentiation of more mature phloem tissue, and appears in the form of active prosenchymatous cells, lacking companion cells and supported by parenchyma (Fig. 5·2).

The sieve elements of mature metaphloem are enucleate and show cytoplasmic continuity with their neighbours across the

30

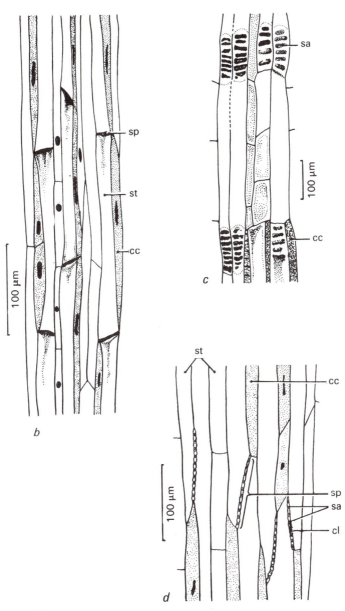

See opposite page.

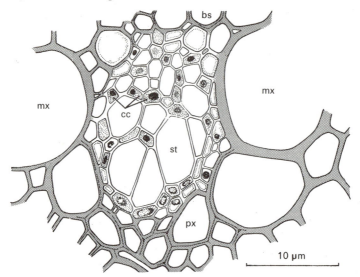

Fig 3·7 T.S. of part of a vascular bundle from the stem of *Asparagus sp.*

micro-pores which penetrate the so-called *sieve areas*. Sometimes the latter are confined to an end-wall forming a *sieve plate* (Fig. 3 · 6).

The companion cells that characterise the phloem of flowering plants have a prominent nucleus, and may control the activity of the neighbouring sieve elements. They are absent from gymnosperms.

4. EPIDERMAL CELLS

These vary in length and shape according to species, and according to their position on the plant. The epidermis is formed of a close mosaic of cells without intercellular spaces. Their outer tangential and often their radial walls are cutinised as well as lignified, and they nearly always show an excreted cuticle, lying as a waxy layer on top of the wax-impregnated and cutinised wall (see Figs. 1 · 1 and 3 · 8). The epidermis presents to the exterior an impervious and non-wettable surface. It restricts non-stomatal water loss, helps to prevent the entry of pathogens, and forms a taut coherent external envelope which by resisting the turgor expansion of the parenchymatous cells within helps to maintain the rigidity of the plant (see 1 · 2 : 3). The stomatal apparatus constitutes a special development of the epidermis and will be considered more fully in chapter 5 · 5.

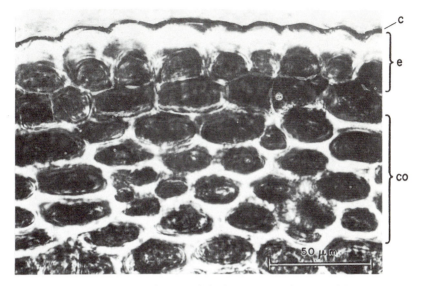

Fig 3·8 T.S. of the outer stem of *Tilia sp.* in its first year, showing the cuticle (c), the epidermis (e), with heavily thickened and cutinised outer wall, and abundant sub-epidermal collenchyma (co). Compare this with Fig 4·13 showing a later stage in the development of periderm. [Ph.]

5. MERISTEM CELLS

These are very thin-walled parenchymatous cells of theoretically unlimited growth. They are located at the growing points of stem and root (apical meristems), at the edges of developing leaf rudiments (marginal meristems), or between primary xylem and phloem cells in the so-called *lateral meristems* (which include vascular and cork cambia). Increase in length is brought about in some plants (e.g. at grass nodes, the base of daffodil leaves etc.) by the activity of *intercalary meristems*.

4 The main patterns of tissue arrangement in higher plants

4 · 1 *Introduction*

Tissues are arranged in stems, roots and leaves in fairly easily distinguishable patterns, though, as might be expected, there are many variations upon the few representative themes to be shown here. In the course of evolutionary time, the flowering plants have shown a separation into two main streams—the monocotyledons (with one seed leaf) and the dicotyledons (with two seed leaves). The former are for the most part an herbaceous group, though the palms are tree-like, whilst the dicotyledons include herbs, shrubs and trees. The gymnosperms, including the conifers, are an essentially woody group.

Woodiness depends upon the degree of development of lateral meristems (the vascular cambia) whose activity in woody plants follows hard upon the completion of differentiation of the primary xylem and phloem tissues. Whether in stem or in root the lateral meristems contribute *only* to increase in girth; all extension growth will already have been accomplished by normal extension of primary tissues. Only after the completion of primary extension do stems and roots increase in girth by means of their vascular cambia. The development of woody cambia is not observed at all in monocotyledons except as an occasional anomaly.

It is proposed here only to illustrate the main features of adult stem structure as found in (*i*) relatively unthickened herbaceous dicotyledons, (*ii*) woody herbaceous dicotyledons, (*iii*) woody shrubs and trees and (*iv*) herbaceous monocotyledons. Monocotyledonous and dicotyledonous roots will also be compared. A section will be given to the mode of action of lateral meristems, including cork formation. The remaining part of the chapter will deal with leaf structure in its broader aspects, though certain aspects of leaf structure will be developed more fully in Chapter 5 on photosynthesis and gas exchange. Reference should be made to the appropriate figures whilst reading the following account.

*General comparison of root and stem structure**

It should be noted that all primary vascular tissues in stems are peripheral in position, whilst in roots the primary vascular tissues are seen to be central (Compare Figs. 4 · 2 and 4 · 7). This follows a good engineering principle, viz. that if a builder or constructional engineer is making provision for tension or thrust he uses a solid rod-like structure (as in a tie-rod which opposes the outward movement of the walls of a house); if however he is making provision against bending strains (as in an aerial mast) he uses an open lattice work structure.

In the vascular bundles of most stems, phloem tissues (stippled in the diagrams) lie to the outside of the woody xylem and on the same radius, giving what are known as collateral bundles (Fig. 4 · 4). In families such as the Solanaceae (tomatoes, potatoes) and the Cucurbitaceae (marrows and gourds), groups of phloem are also found internal to the xylem and the vascular bundles here are described as bicollateral (Fig. 4 · 1).

In the root, phloem groups lie in positions that alternate with the xylem arms, i.e. they are on alternate radii (xylem is radially shaded in diagrams). The first differentiated xylem (*protoxylem* is shaded black) lies to the inside of each bundle in a stem (in the *endarch* position) and later differentiated tissues develop to the outside of it in a *centrifugal* direction (i.e. away from the centre). In a root the protoxylem groups are in the *exarch* position, and differentiation is said to be *centripetal* (i.e. towards the centre) (Fig. 4 · 7).

4 · 2 The structures of typical stems and roots

4 · 2 : 1 Herbaceous dicotyledon stem with little secondary thickening
Examples: *Stellaria media* (chickweed) and *Anthriscus sylvestris* (cow parsley); see Fig. 4 · 2.

Note the relatively distinct vascular bundles, well separated by broad primary rays which connect a broad pith to a narrow cortex. The cortex may be partly photosynthetic. A cambium develops but does not contribute much in the way of secondary tissue. In some

---

* For further details on these and other anatomical matters consult Esau's *Anatomy of Seed Plants* (see Further Reading at end).

families (such as the Ranunculaceae) the absence of cambium and the tendency to scattering of the vascular bundles is thought to provide evidence for the common origins of monocots and dicots. (See figure 4 · 6).

4 · 2 : 2 Herbaceous dicotyledon stem with secondary thickening
Examples: *Dahlia* and *Vicia faba* (broad bean); see Figs. 4 · 3 and 4 · 4b.

There is often a greater tendency in plants of this type to develop a good deal of supporting sclerenchymatous tissue. Many herbaceous perennials die back in the fall when the autumn frosts strike them. Despite their tenderness, they may show well-developed cambial activity and may have produced much woody tissue by the end of the season.

4 · 2 : 3 Woody dicotyledon stem
Example: *Fraxinus* (ash); see Fig. 4 · 7a, b and c.

Figure 4 · 7a shows a section taken very early in the season; there are a number of closely positioned narrow bundles, separated only by narrow parenchymatous primary rays. Cambial activity quickly starts to add secondary wood and phloem, and at the end of the first season (Fig. 4 · 7b) most of the ring of woody tissue is seen to be secondary. After a second season (Fig. 4 · 7c) it is possible to distinguish between the two successive growth increments because the elements laid down at the end of the first season (summer wood) are smaller than those which are laid down in the spring of the second season (spring wood). This can be related also to the larger number and diameter of the water-carrying vessels that are laid down in the spring wood, at a time when the water demand is high and when vegetative growth is in full flush. A larger proportion of supporting fibres is found in the summer wood.

4 · 2 : 4 Herbaceous monocotyledon stem
Example: *Zea mais* (sweet corn or maize); see Fig. 4 · 5.

Note the scattered individual bundles; these are *closed*, that is they do not develop a vascular cambium, and they are generally invested with a sclerenchymatous bundle sheath. There is a tendency in spite of the scattering for the concentration of bundles just beneath the epidermis, i.e. at the stem periphery, and there is often a marked development of fibres amongst the subepidermal tissues.

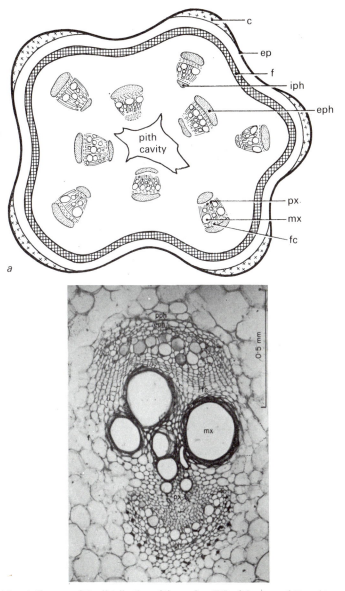

Fig 4·1a. A diagram of the distribution of tissues in a T.S. of the stem of *Cucurbita sp.*, showing bicollateral bundles with internal and external phloem, and a pith cavity, surrounded by a continuous ring of perivascular fibres (f), and ridges of collenchyma (c).

b. Photograph of a whole bundle from a T.S. of the stem of *Cucurbita sp.* at higher magnification, showing protophloem (pph), external phloem (eph), internal phloem (iph), metaxylem (mx), protoxylem (px), and fascicular cambium (fc).

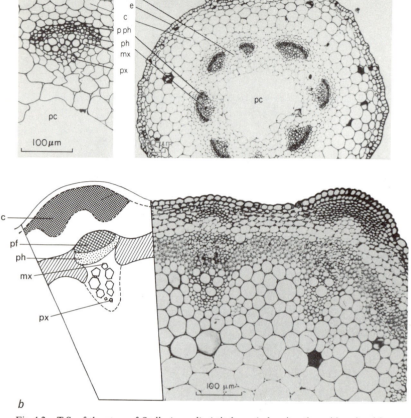

b

Fig 4·2a. T.S. of the stem of *Stellaria media* (stitchwort) showing the epidermis with multicellular hairs. Broad primary rays (pr) separate the vascular bundles and connect pith and cortex, and there is a pith cavity. The enlarged view of a collateral vascular bundle shows xylem to the inside and phloem to the outside, and the earliest differentiated phloem is beginning to collapse. Labels are as before, including pith cavity (pc) and cortex (cl). (L.G.B.)

b. T.S. of the stem of *Anthriscus sp.* (beaked parsley) showing groups of collenchyma cells which occupy the ridges and correspond in position to the vascular bundles. Each bundle consists of protoxylem (px), metaxylem (mx), and phloem and phloem fibres to the outside (ph and f). Between the bundles, fibres occupy the outermost part of the stele and interrupt the primary rays which would otherwise link pith to cortex (cf. *Dahlia sp.* Fig. 4·4a). (L.G.B.)

Note the tendency to flattening of the outermost bundles, which presumably reflects the restraint exercised by the tightly coherent sheath of epidermal cells on the expanding tissues within. Nearer to the centre the bundles are radially elongated in a more normal manner, and they are under less stress at this point. It should be

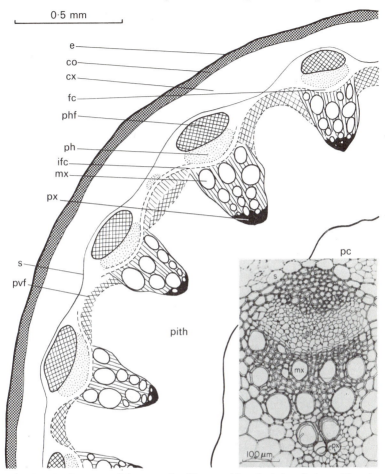

Fig 4·3 T.S. of a sector of the stem of *Dahlia sp.*, with a single bundle inset. There is an almost continuous ring of sub-epidermal collenchyma. A starch sheath (s) represents the inner limit of the cortex and may later give rise to a primary endodermis. Note also the difference between the phloem fibres (f) and the perivascular fibres in the interfascicular position (pvf).

remarked also that in monocotyledons, since there are no lateral meristems, stem girth can only increase by the lateral extension of cells or by bulk increase in the size of the apical meristem. It can often be observed that extension growth leads to a rupture of the protoxylem tissues, leaving a protoxylem cavity with fragments of torn vessel thickenings clinging to its margin.

39

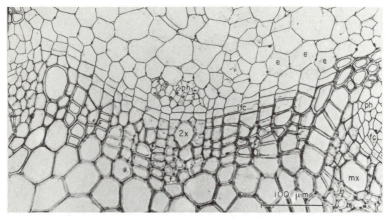

a

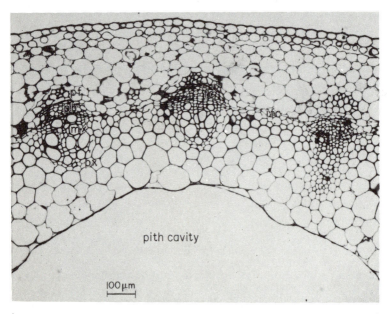

pith cavity

100 μm

b

Fig 4·4a. T.S. of part of the stem of *Dahlia sp.* showing the extension of fascicular cambium (fc) towards an adjacent bundle. An endodermal layer (e) is indicated by the presence of Casparian bands (see page 101).
b. T.S. of a part of the stem of *Vicia faba* (broad bean) showing the origin of the fascicular cambium and its extension (ifc) to join up with that from an adjacent bundle. (L.G.B.)

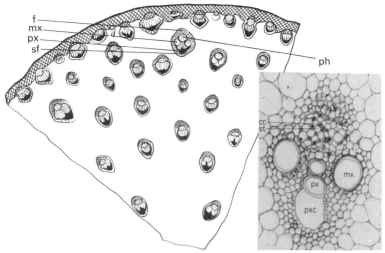

Fig 4·5 A monocotyledon stem.

A sector of a T.S. of the stem of *Zea mais* (maize) and (inset) one of the vascular bundles. The bundles are scattered, and those at the periphery tend to be flattened radially and are associated with a good deal of fibrous development.

The bundle shown is from a young stem, and the phloem fibres (f) and bundle sheath fibres (sf) are as yet undifferentiated. Rapid growth has torn the developing protoxylem, leaving a protoxylem canal (pxc). There are two large metaxylem vessels (mx); in some monocotyledonous bundles the phloem (ph) is deeply sunken between the metaxylem vessels. The phloem tissue is very regular, because there is little phloem parenchyma, and sieve tubes (st) and companion cells (cc) form a regular pattern.

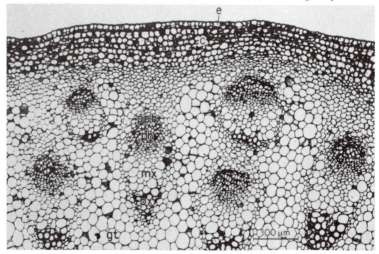

Fig 4·6 T.S. of a portion of a young stem of *Thalictrum sp.* (rue) showing scattered bundles, and ground tissue (gt), with other labels as in Fig 4·1. It is difficult to distinguish between pith and cortex (cf. monocotyledon stem, Fig 4·5).

41

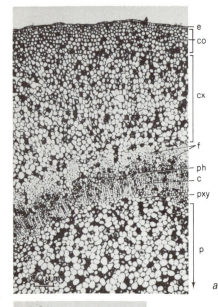

e
co

cx

f
ph
c
pxy

p

100 μm

a

Fig 4·7 Woody dicotyledon stems.
a. T.S. of a very young stem of *Fraxinus excelsior* (ash). At this stage the pith (not shown in full) is about 2 mm in diameter, and the cortex is between 0·5 mm and 0·7 mm wide. The vascular cylinder is only about 0·25 mm wide. The phloem fibres (f) are still thin-walled, and show signs of collapse, due to the epoxy-resin embedding technique employed. The cambial ring is already complete and has added some secondary xylem to the cylinder. (L.G.B.)

See opposite page for captions to *b*, *c*, *d*.

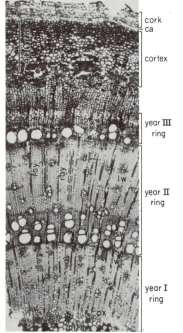

cork
ca

cortex

year **III** ring

year **II** ring

year **I** ring

c

50 μm

d

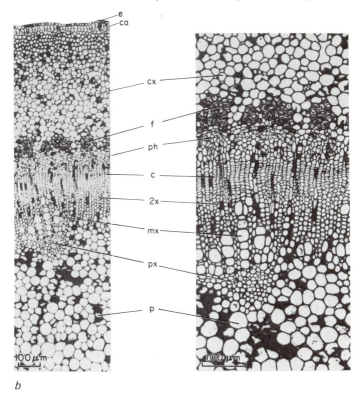

b

b. T.S. of an ash stem towards the end of its first year of growth. By now a cork cambium (ca) has differentiated below the epidermis. The cortex remains 0·5 mm wide, but is showing some sign of disruption. The vascular ring has doubled its width, but few vessels are added at this late stage of the year. The phloem fibres are by now thick-walled, and a few primary rays, 1–2 cells in width, pass from pith to cortex. (L.G.B.)

c. T.S. of a stem of ash after 2·5 years' growth. The vascular ring is now about 2·0 mm thick, and as it has expanded much of the inner cortex has been torn or compressed. A periderm is now fully developed, and the original epidermis has been shed. Such a stem is described as ring-porous. The limit of each annual increment of wood is marked by the line separating the ring of large spring vessels (sw) from the preceding late-summer contribution of small fibres (lw). (L.G.B.)

d. T.S. of a 4-year old stem of *Tilia europaea* (lime). An active cambium (c) has given rise to secondary xylem (2x) on the inside, and alternating bands of phloem (ph) and phloem fibres (f) outwards, successive increments being numbered ph_1, ph_2, etc. Uniseriate rays (rf) run from the xylem across the cambium and can be traced out through the sieve tubes (st) and companion cells (cc). One ray ends in the characteristic funnel-shaped expansion (v) separating neighbouring wedges of phloem. (L.G.B.)

43

4 · 2 : 5 Primary root: dicotyledon vs. monocotyledon

Examples: *Vicia faba* and *Zea mais*.

Figures 4 · 8 and 4 · 9 show the essential differences in tissue arrangement between dicot and monocot roots, and they also underline the differences between root and stem already mentioned. Roots are absorbing, anchoring, supporting and sometimes storing organs. As will be discussed in Chapter 7 their epidermal absorptive regions become replaced by a suberised exodermis. Where roots develop secondary thickening, the exodermis and much of the tissue external to the phloem is shed after the establishment of a protective periderm (see Section 4 · 3 : 3), formed from another type of lateral meristem.

Unthickened dicot roots have a central woody stele, generally with from two to five protoxylem groups; roots are described as diarch, triarch, tetrarch or pentarch according to the number of protoxylems present. On the other hand, monocotyledon roots generally have a larger number of protoxylem groups and are described as polyarch. Phloem tissue is seen to lie in the space between adjacent protoxylem groups and on alternating radii. The small amount of tissue between xylem, phloem and endodermis is called pericycle. This region of thin-walled parenchyma is the site of origin of lateral roots (Fig. 4 · 10c) which, because they arise from a submerged position in the root tissue, are said to be *endogenous* in origin. Lateral roots have to grow across the cortex and burst through the tissues to the exterior. (Adventitious shoots may sometimes also be endogenous in origin, but normal buds on a stem are of superficial or *exogenous* origin.)

The *endodermis*, a single layered sheath of cells around the vascular stele, is constructed so that the radial transport of water and solutes across the sheath is likely to be limited to the protoplasts of its constituent cells. The endodermis is described in greater detail in Chapter 5.

In unthickened roots the broad parenchymatous cortex may store reserves of starch etc. but it may be more important to the root as a transport route for gases, which can diffuse through a relatively extensive intercellular space system.

The last-differentiated more central cells of the vascular stele do not always give rise to vessels, especially in monocotyledons where they may form a narrow lignified pith, or sometimes an even broader non-lignified pith (as in *Zea mais*).

44

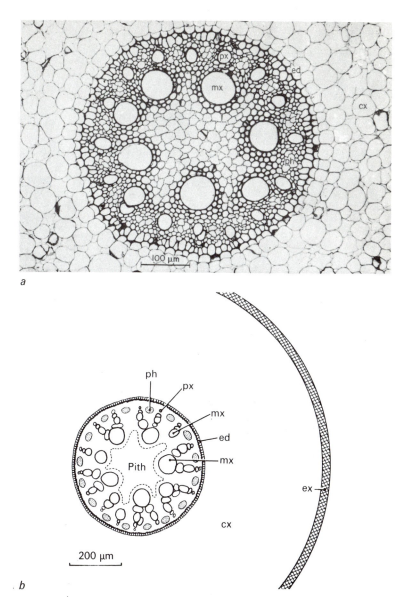

Fig 4·8 The diagram shows the structure of a root of *Zea mais* (maize), a monocotyledon, in T.S. The inset photograph shows details of the polyarch stele, in this case with 16 exarch groups of protoxylem, and phloem islands on alternating radii. Large metaxylem vessels (mx) seem to be related to one, two, or at most three protoxylem groups. The pith consist of unlignified parenchyma, which may later become lignified. The endodermis (ed) is of tertiary form, and the cortex (cx) is well provided with an air-space system. The exodermis (ex) is prominent.

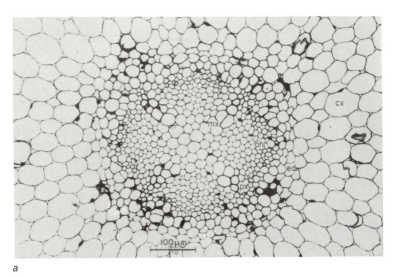

a

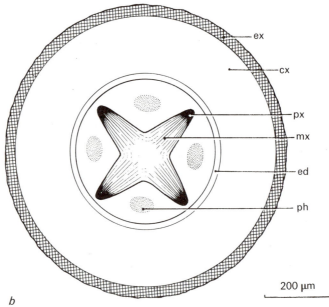

b

Fig 4·9 T.S. of the root of *Vicia faba* (broad bean) which is a dicotyledon. The diagram indicates the broad cortex (cx) and the central stele, which is tetrarch. The four exarch protoxylem groups (px) alternate with four phloem groups (ph). The central tissues are not yet differentiated but will later lignify, and some of the central elements may differentiate to vessels. The primary endodermis (ed) is not yet clearly defined, but the abundant air-spaces of the relatively broad cortex (cx) can be seen. There is again a marked exodermis (ex). (L.G.B.)

46

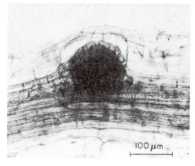

a

b

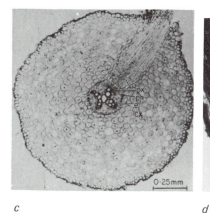

c   *d*

Fig 4·10 *a*, *b*, and *c* Lateral roots.
Photographs showing stages in the emergence of lateral roots from the main root of *Polygonum persicaria* (spotted persicaria). The lateral primordium arises in the pericycle just within the endodermis, and grows outwards through the cortex, bursting through to the outside as a root with a fully formed apical meristem.

The roots in Figs. 4·10*a* and *b* were fixed in glutaraldehyde fixative, and cleared in lactophenol-acid fuchsin. That in 4·10*c* was mounted and photographed in water to show the extensive air space system (a, black lines) reaching almost to the apex of the lateral root.

d. T.S. of the root of *Ranunculus repens* (creeping buttercup) showing a lateral root which has originated in the pericycle (p) just within the endodermis (e), has grown across the cortex (cx) and broken through the surface tissues.

4 · 3 *The problem of increase in size*

The colonisation of land surfaces by green plants presumably involved two linked processes; one was the improvement which could now be brought about in the mechanics of the photosynthetic process; this led to the second, viz. the need for increased control of water loss. The leaf, as it has evolved, offers a very effective compromise solution of the problems involved, and the net result is that much more capital is made available for growth. The fossil evidence suggests that the woody tree-habit was prevalent from very early times, and it seems likely that once the problem of continued (or renewed) meristematic activity had been solved (with the production of lateral meristems) plants would have been able to support a much larger leaf canopy, with the aid of the increased mechanical and conducting tissues available as a result of secondary growth. A larger photosynthetic output means more food for growth and reproduction; it means a greater potential for longevity and survival.

Size carries with it these advantages but also certain obvious disadvantages. The giant redwoods and the tall blue gums may dominate their respective communities in the competition for light but, for example, the redwood may have to raise water through nearly 100 metres in order to maintain its leaf canopy in good working order. Wind velocities increase with total height above the ground and it seems likely that for this reason alone mechanical strains will be more hazardous and water loss increased. To achieve a larger safety margin a massive capital has to be laid down in the form of ligno-cellulose, which is literally a dead loss to the plant for the purposes of subsequent growth.

In brief, secondary thickening is useful only up to certain limits, beyond these limits mere size increase might well prove less advantageous. Nevertheless it is pertinent that we should briefly examine the mechanism of secondary thickening in the stem and root of higher plants since it is of such widespread occurrence amongst the dicotyledons and the gymnosperms.

4 · 3 : 1 Secondary growth in the dicotyledon stem

Figures 4 · 2a, 4 · 2b show typical herbaceous dicot stems at a stage before secondary activity has occurred. Compare them with Fig. 4 · 4. The cambium, which remained as a sheath of undifferentiated primary tissue lying between the most recently differentiated xylem (to the inside) and phloem (to the outside), now begins to display cell divisions of a characteristic pattern. In herbaceous species which develop secondary thickening, activity starts within the vascular bundles (*fascicular cambium*) and extends across the intervening ground tissue (primary ray) to join up with the developing cambium from a neighbouring bundle (Fig. 4 · 4b). Where the cambium bridges the gap between adjacent bundles it is known as *interfascicular*. By its further activity, the cambium develops to form a continuous sheath, interpolating a cylinder of secondary tissue between the already established primary xylem and phloem. In woody twigs (e.g. Fig. 4 · 7) the rays are so narrow and the primary bundles so close that it is inappropriate to distinguish fascicular and inter-fascicular cambium. Secondary growth by cambial activity quickly follows primary tissue production and differentiation, as woody tissues are added to the lengthening first-year twig. A very similar state of affairs is characteristic of gymnosperm stems.

The cells which make up much of the cambial sheath are described as fusiform, because in longisection they appear to be spindle-shaped with more or less tapering ends (Fig. 4 · 11). Mitosis takes place and a cell wall is laid down in a tangential plane (i.e. perpendicular to a radial line through the stem which joins pith and cortex). The cells cut off in this way are called *cambial initials*; if they are cut off to the inside (x) of the cambial sheath (c) they differentiate as cells of the xylem, i.e. as vessel elements, fibres, etc.; if cut off to the outside (p) they contribute to the phloem. The sheath is interrupted at intervals where ray tissue penetrates the wood and continues outwards into the phloem, and the fusiform initials at this point are replaced by ray initials which are shorter and much more cuboid in shape, giving rise to parenchymatous cells which add on to the rays that run horizontally through the wood and phloem. Ray tissue assumes greater importance in woody shrubs and trees, for not only does it constitute a radial and horizontal translocation pathway which traverses the otherwise mostly dead wood, but it also can serve as a storage tissue, holding reserves which may be important, for example during the spring flush of growth.

It will be seen that since the cambium comes to form a complete sheath around the woody stele (in root as well as in stem; see Fig. 4 · 12) the increments of tissue that it adds year by year, the so-called

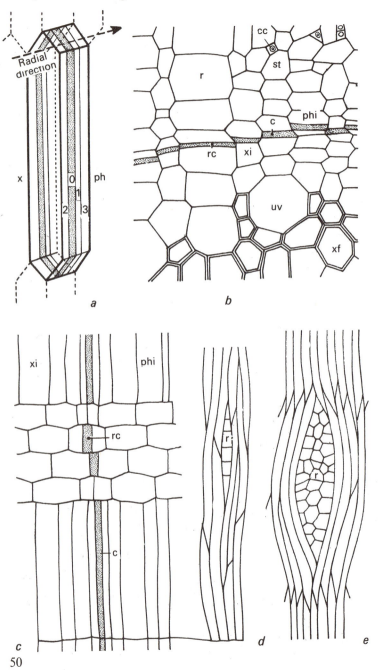

annual rings, virtually form successive cones of secondary tissue extending from the ultimate branches of the tree downwards into the tree trunk (and so on downwards into the root system). The break of dormancy in the buds leads to a renewal of hormone-stimulated cambial activity in the branch tips, and this spreads downwards through the whole plant, sparking off the production of the current annual increment of secondary xylem and phloem, and ultimately reaching the extending primary root system. It may take something of the order of three weeks or more in a normal season before the activity which started in the growing buds of a tree has extended right through to those parts of the root system which will undergo secondary thickening.

4 · 3 : 2 Secondary growth in the dicotyledon root

Figures 4 · 12a and b show the process of cambial development during the start of secondary growth in dicot root. It will be seen that cambial activity starts within a very narrow zone of undifferentiated parenchyma between phloem and xylem and extends towards the protoxylem arms. Rapid differential growth (cutting off more xylem than phloem) makes it possible for a fairly regular ring of cambium to be established; opposite to the xylem arms, ray tissue is cut off which establishes a link between the primary wood and the tissues to the outside.

Fig 4·11 Cambium.

a. Diagram showing the products of a simplified cambial cell. Two tangential walls have isolated a new cambial cell (0) from an outer (1) and an inner (2) cambial initial; a third wall has then cut off a new initial (3) to the outer (cortical) side of the cambium. (1) and (3) will become secondary phloem tissue, and (2) will differentiate as a secondary xylem cell.

b. A diagrammatic representation of cambium of a woody plant in T.S., cutting off phloem tissues to the outside and secondary xylem to the inside.

c. Diagram of R.L.S. through a woody stem, showing that where a ray passes horizontally through the stem it traverses xylem and phloem and is itself increased by cambial activity.

d and *e.* Diagram of a stem in T.L.S. (tangential longitudinal section) showing a cross-sectional view of a ray in the position at which it crosses the cambium. In *d* the ray is uniseriate and the cambium is non-storeyed. In *e* the ray is multiseriate and the cambium is storeyed, i.e. all the fusiform cambial cells are of the same order of length, and all end at about the same level.

Abbreviations: r, ray; rc, ray cambium; cc, companion cell; st, sieve tube; phi, phloem initial; c, cambium; xi, xylem initial; uv, undifferentiated vessel; xf, xylem fibre.

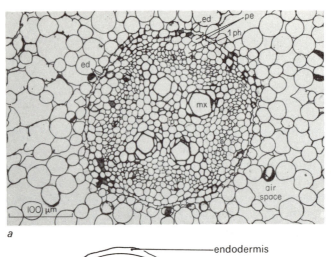

a

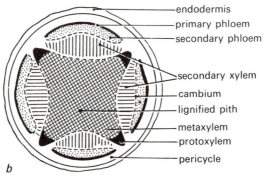

endodermis
primary phloem
secondary phloem

secondary xylem
cambium
lignified pith
metaxylem
protoxylem
pericycle

b

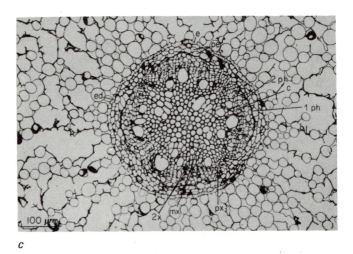

c

52

4·3:3 **Periderm**

The very fact that increase in girth is taking place in stem or root as the result of the action of a woody cambium means that there is likely to be increasing strain on the exterior tissues. The epidermis in the stem and the exodermis in the root (which replaces the withered root hair layer) are both coherent structures lacking intercellular spaces. As new tissues are added internally it is inevitable that for a time cortical tissues will be compressed against the containing outer sheath of cells, and tangential strains will increase. However, before the break comes, certain cells of the cortex undergo activation to form what eventually becomes a continuous cambial sheath known as a *phellogen* or cork cambium. This is notable mainly for the cells that it cuts off to the outside, for these cells rapidly expand and after a period of fatty metabolism during which suberin is progressively laid down in their walls, they die because they are cut off from their sources of supply. In the process they come to form an impermeable covering which markedly restricts water loss and helps the plant to resist invasion by pathogens. Sometimes the phellogen produces to the inside what is effectively an addition of parenchymatous cells to the cortex; it is known as *phelloderm* and the whole protective sheath comprising cork cambium or phellogen and its products is known as *periderm*. (Fig. 4 · 13).

Periderm may arise quite superficially in some stems, e.g. in *Salix* (willow) and in *Populus* (poplar) it arises just beneath the epidermis; in other plants it is much more deeply seated. Tissues external to it are isolated from their nutritional sources and are ultimately sloughed off. In trees which show extensive cork formation, the bark often develops fissures, or it may peel off in scales in quite characteristic fashion. Similarly in roots periderm arises either superficially (especially where there is little secondary

---

Fig 4·12 Secondary thickening in a dicotyledonous root.

a. T.S. of a tetrarch root of *Salix sp.* (willow) in which cambial activity has just been initiated between xylem and phloem. Primary phloem groups (1ph) lie on alternate radii to the four protoxylem groups (px) within the endodermis (ed).

b and *c.* Diagram and T.S. of an older root of *Salix sp.* The cambium (c) has not yet completed a continuous ring round the xylem but will eventually do so. Secondary wood is hardly distinguishable from primary wood, and secondary xylem elements (2x) have been differentiated so as to be practically continuous with the primary stele. Primary phloem has been crushed or obliterated by the secondary growth (2ph) though the pericycle (pe) has withstood the compression. The key to the position of the exarch protoxylem is to spot the discontinuities between the lines of obliterated primary phloem, which lie on alternate radii to the protoxylem arms. Lignified pith can be seen (lp). (All L.G.B.)

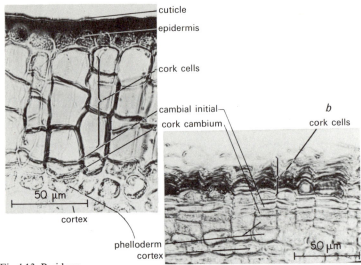

Fig 4·13 Periderm.

a. T.S. of the outer stem layer of *Populus sp.* (poplar). A cambium has cut off three or four layers of cork cells to the outside of the stem. Only one layer of phelloderm cells has been cut off adjacent to the cortex. The epidermis with its cuticle and heavily cutinised outer layer is still intact.

b. T.S. of the outer layers of the stem of *Tilia sp.* (cf. Fig 3·8 showing an earlier stage). Here the epidermis has been cast off and the irregular ranks of cork cells can each be related to a cambial cell. Two phelloderm cells are seen beneath each cambial cell.

vascular tissue formed and the cortex is allowed to remain), or in a more deep-seated position.

In the case of the roots of dicots or of conifers, the periderm is usually of fairly deep-seated origin, but otherwise develops as already described. In monocotyledons, where most often the root system is adventitious, and the primary system is of very short duration, there is very little cork-formation in most species. The coconut palm shows a root periderm like that in dicotyledons; in a few other cases corky cells are derived from deep-seated cortical parenchyma cells but no regular phellogen is formed.

The total sheathing of a root or stem with an impenetrable corky barrier does raise problems of ventilation and it is common to find that the cork is interrupted at intervals by pores known as *lenticels*. At such points the phellogen actively cuts off a loose tissue with a considerable development of intercellular spaces; it is known as *complementary* tissue and its constituent cells may be corky (as in pear and apple) or non-corky (as in ash or oak).

4 · 4 *The leaf*

Examples: *Ligustrum sp.* (privet) (Figs. 4 · 14 and 4 · 15) and *Dactylis sp.* (cocksfoot grass) (Fig. 4 · 17).

The leaf serves to display the photosynthetic apparatus to its best advantage. The chloroplasts in the cells of the mesophyll need to be adequately supplied with water and carbon dioxide. The protective epidermal layers and the skeletal vascular system all combine to regulate water content, gas diffusion and the removal of photosynthetic products.

Characteristically the leaf is a dorsiventrally flattened structure and needs to be investigated in at least three planes in order to comprehend its structure fully. Figures 4 · 14a to d show the result of cutting surface and paradermal sections (i.e. parallel to the dorsiventral plane). Surface sections show either upper or lower surfaces (Fig. 4 · 14a and b); in effect one is looking at and through the epidermal cells, either from the outside or the inside of the leaf, and stomata are seen in *optical* section. Such sections also make it possible to see the relationship between the epidermal cells and stomatal pores and the mesophyll and airspaces beneath. In paradermal sections, an upper plane of section would traverse the palisade tissue (Fig. 4 · 14c), a mid-plane would show the vascular tissue and associated mesophyll cells, whilst a lower plane would mostly traverse spongy mesophyll tissue and its abundant airspace system (Fig. 4 · 14d; see also Figs. 5 · 2a and b). The main advantage of a longitudinal vertical section is to show alternative views of stomata; this of course will in any case depend on how they are oriented in the epidermis. It may also reveal airspace systems that are sometimes concealed when dealing only with transverse vertical sections, e.g. in certain conifer leaves such as *Pinus*, where plates of mesophyll cells alternate along the length of the leaf with the airspaces that are sandwiched between them.

Attention may be directed to the following points. The main vein or midrib is a continuation of the petiole or leafstalk. This happens to be short in privet but may be a good deal longer in other plants. The petiole may have a vascular structure that is stem-like, or in view of its need to accommodate to the movements of a leaf lamina that bends and twists in the wind, the vascular strands may be arranged like an open C or an inverted omega or horseshoe (Ω) (Fig. 4 · 16). The midrib may show secondary thickening, especially in woody plants; since a leaf bundle is continuous with

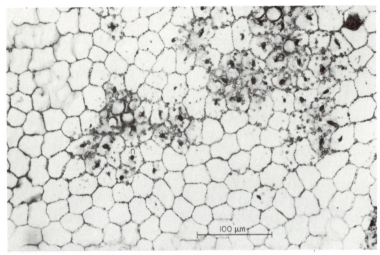

a

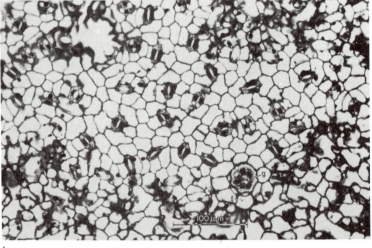

b

Fig 4·14 Leaf of *Ligustrum vulgare* (privet).
a. Horizontal section (H.S.) through the upper epidermis, showing the conspicuous dark nuclei. The granules are not chloroplasts, which are rarely found in epidermal cells. This is a close-knit tissue with no air spaces. There is some evidence of pitting in the anticlinal walls (perpendicular to the surface).
b. Horizontal section through the lower epidermis of a leaf, a similarly continuous 'skin' tissue, with open stomata and one multicellular gland (g).

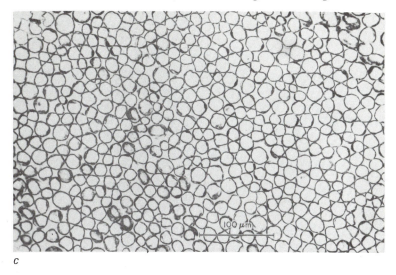

c

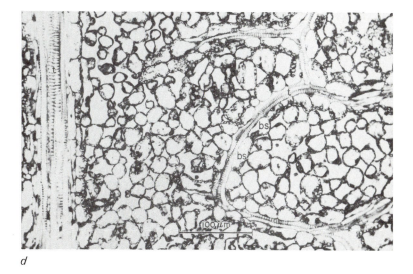

d

Fig 4·14 continued.

c. Paradermal section (P.S., parallel to the epidermis) through the palisade mesophyll (cf. Fig 4·15), showing air spaces and chloroplasts.

d. P.S. through the upper spongy mesophyll and veins, showing tracheids (t) and bundle sheath cells (bs). Air spaces are more abundant than in 4·14*c* and the whole tissue is more irregular. (All L.G.B.)

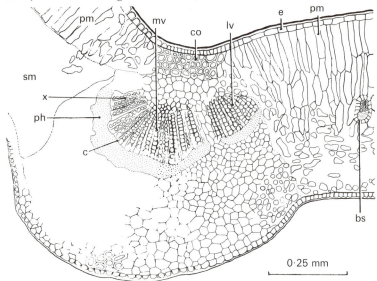

Fig 4·15 A semi-diagrammatic drawing of a T.S. of the leaf of *Ligustrum vulgare*, showing the main vein (mv), from which a lateral vein (lv) has just parted. Xylem is found towards the upper (adaxial) side of the leaf, whilst phloem is found on the lower (abaxial) side. Cambial growth (pm) has started, and collenchyma (co) can be seen above the vascular bundle. The palisade mesophyll (pm) is two- or three-layered, and the cells are columnar in shape (cf. Fig. 4·14c). Spongy mesophyll (sm) and a small vein enclosed in bundle sheath cells (bs) are shown (cf. Fig. 4·14d).

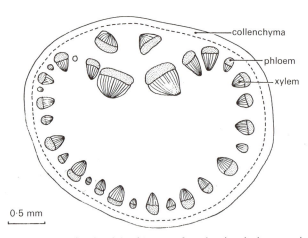

Fig 4·16 Diagram of the leaf petiole of *Vitis vinifera*, showing the inverse orientation of the two topmost bundles; the vascular bundles show an inverted omega-shaped distribution (℧).

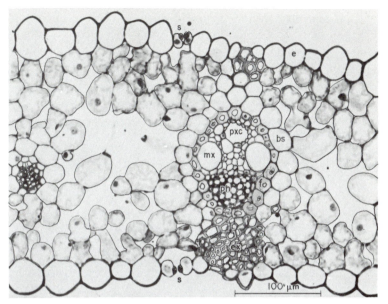

Fig 4·17 T.S. of a leaf of *Dactylis glomerata* (cocksfoot grass). The large vascular bundle shows a bundle sheath (bs) and bundle sheath fibres (f), with a bundle sheath extension on the abaxial side (ex). The phloem is also on the abaxial side; it is heavily stained and quite thick-walled. A typical protoxylem canal (pxc) and the two metaxylem vessels (mx) contribute to the appearance of the 'monkey-face' bundle. No chloroplasts are found in the epidermis (e), and stomata (s) may be found on both sides of the leaf; the guard cells are flanked by larger subsidiary cells. There is no obvious distinction between palisade and spongy mesophyll.

the vasculation of the stem, it is not difficult to remember that the xylem, with protoxylem uppermost, will lie towards the upper *adaxial* surface of the leaf, whilst the phloem will lie towards the lower abaxial surface (Fig. 4 · 15). The main vein and to a lesser extent the subsidiary veins may be substantially backed with collenchymatous tissue generally on the dorsal side; this tissue has already been described (pp. 22 and 25) as retaining flexibility whilst providing good support. The net veined leaves of most dicotyledons may be contrasted with the parallel (or striate) venation of monocotyledons (though there are often cross connections which link up these parallel veins). There are other minor points of contrast. Often the mesophyll of monocotyledons is less well differentiated into palisade and spongy tissues, and apart from the orientation of xylem and phloem there is no very real distinction between upper and lower surfaces (Fig. 4 · 17). Monocotyledon leaves frequently have basal intercalary meristems, and (as in daffodil, for example) grow in length by tissues which are interpolated at the leaf base.

4 · 5 *Modifications of stem, root and leaf*

It is inappropriate to spend time here on variations from the normal form of the organs that we have described. Whenever it happens, modification is for the most part related to water, either in deficit (xerophytism) or excess (hydrophytism), or to the need for food storage. One may briefly recall, for example, that water-plants (or hydrophytes) have cuticles that are thin or negligible; such plants may show greatly reduced vasculation and an abundant airspace system. Xerophytes, on the other hand, may show a reduction of their surface/volume ratio, and increase in cuticle thickness, sunken and protected stomata, a diminished airspace system and so on.

Storage may occur in stems, leaves or roots, and necessitates some proliferation of storage parenchyma, often at the expense of other tissues. Examples of storage are well enough known, e.g. (*a*) swollen root tissues in carrot (sugars), *Dahlia* (inulin), cassava (starch); (*b*) swollen stem tissues in potato tuber (starch), various rhizomes such as *Iris* (irisin—a fructosan like inulin), Jerusalem artichoke (inulin); (*c*) in leaves (e.g. scale leaves of tulip bulb) a swollen bud which stores starch; the leaf bases in wood sorrel (*Oxalis*) store starch.

In examining and recognising the morphological nature of storage organs of this kind, one must be guided by general morphological precepts, including the following:
1. A root normally only gives rise to roots like itself.
2. A stem bears leaves (sometimes reduced to scale leaves), buds and perhaps adventitious roots. The latter can be seen to break the surface of the stem at their point of exit.
3. Leaves (including scale-leaves) are generally found to subtend a bud in their axil, i.e. in the angle at the junction of leaf and stem.

PART TWO

Structure and function in plants

5 Photosynthesis and gas exchange

5 · 1 *Introductory*

In the summary account of some of the molecular aspects of photosynthesis (2 · 2) it was stated that light energy can be used with the help of the chloroplasts of green tissues to build up hexoses and other compounds from the CO_2 of the atmosphere and water. Considering this now from a physiological rather than from a biochemical point of view, it will be seen straightaway that apart from the main internal factors controlling cell metabolism and the intrinsic genetic factors such as those, for example, which control the manufacture of chlorophyll, the following factors are likely to be of major importance in determining the effectiveness of the photosynthetic process:

(*a*) Leaf structure in so far as it mediates the effective supply of water and CO_2 to the green cells, displays them adequately to the light and removes the end-products of the process.

(*b*) External factors, all of them capable of influencing the metabolic processes which are going on in the leaf; these include the quality and quantity of light available, the availability of CO_2 and water, and the effects of temperature.

5 · 2 *Structure and function in the leaf*

Much of what follows in this section can equally be applied to the operation of photosynthesis in organs other than leaves. It must be remembered that stems are often green, and in some plants the bulk of the carbon fixation is brought about by stem tissues whose problems of supply and demand are very similar to those of the better adapted leaf. A good deal of this section is also relevant to the material of Chapter 6 on transpiration. The diffusion of gases into and out of a leaf follow the same principles, and what is relevant

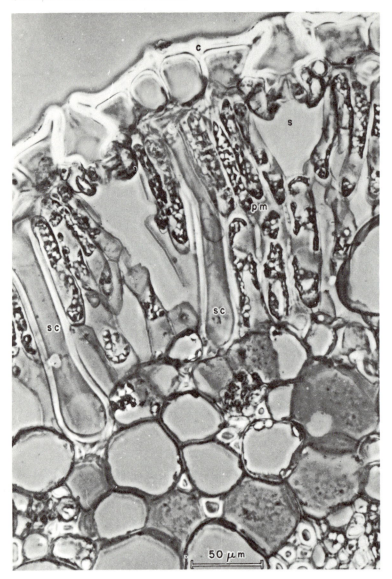

Fig 5·1 A portion of a centric leaf of *Hakea sp.* showing the prominent cuticle (c), deeply sunken stomata (s), double palisade (pm), and sclereids (sc) which aid the leaf against collapse. The central tissue is parenchymatous with vascular bundles which are well provided with fibres. [Ph.]

to CO_2 diffusion inwards is equally relevant to water vapour diffusion outwards.

5 · 2 : 1 Support and transport

The general layout of the leaf has already been discussed in 4 · 4, and only a little more need be said on the mechanical aspects involved. In addition to the supporting framework of petiole, main and lateral veins, some stress should perhaps be laid on the bands of collenchymatous tissues that often give extra support and prominence to the principal veins, as also on the sclerenchyma that sometimes accompanies the smaller veins. The xylem of the veins is responsible for transport of water to the mesophyll and epidermis and it is especially important to bear in mind the importance of turgor in keeping a leaf properly extended. The presence of turgid mesophyll tissues between sheets of epidermal cells tautly stretched over a vascular framework confers an extra rigidity on leaves, the importance of which can best be grasped when wilting sets in (see also Chapter 6).

The leaves of strongly xerophytic plants are often rigid structures (sclerophylls), sometimes reduced to needle form (see Fig. 5 · 1), but in any case they may show such a development of sclerenchymatous supporting tissue that collapse is not possible. Plants like this are called drought-resistant, but they quite often have a high transpiration rate and a deep-rooting system. The important thing is that they do not collapse; by means of a heavy cuticle and sunken stomata they can reduce water loss greatly and some of them can endure quite a high degree of water loss without permanent damage. They may be compared and contrasted with some succulent plants which at most times have a low rate of water loss but which can reduce their water loss practically to zero by the development of massive cuticularisation, and by total stomatal closure.

Leaf veins also have the double function of supplying water and removing the products of photosynthesis; the typically net-veined leaves of dicotyledons have veins that branch into finer and yet finer veinlets, often ending in just one or two tracheids amongst the mesophyll cells. Figure 5 · 2a and b shows something of the relationship of mesophyll cells to transport cells. It is uncommon in most leaves (monocots or dicots) for there to be a distance of more than about 0·1 mm separating adjacent veins, and this ensures efficient transport to and from the mesophyll cells.

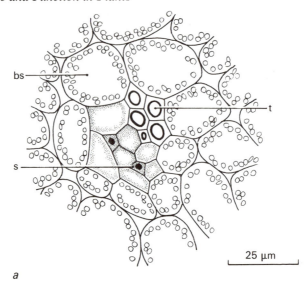

a

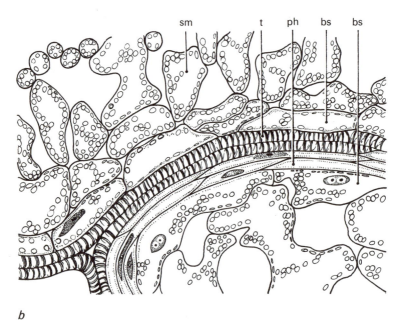

b

Fig 5·2*a* T.S. and *b*. P.S. through the mesophyll of a leaf of *Helleborus sp.*, showing tracheids of xylem (t), phoem (ph) including sieve elements (s) and bundle sheath cells (bs). The spongy mesophyll (sm) has abundant air spaces.

The phloem of these fine veins has few of the characteristics of mature stem phloem. Where present the sieve elements are short and rather narrow and their companion cells are absent or rather difficult to distinguish. Sometimes phloem is represented only by parenchymatous cells at the ends of the veinlets. It is estimated that 99% of the leaf veins are invested in parenchymatous *bundle sheath cells* (Fig. 5 · 2b) and these sheaths are sometimes extended vertically towards or even as far as the overlying or underlying epidermal cells; this gives an increased surface of contact for transport to and from the mesophyll cells. (Fig. 4 · 15). The bundle sheath cells are clearly concerned with the movement of hexose and other manufactured materials away from the mesophyll. They may also be concerned with the conversion of hexose to sucrose which is the main sugar of transport in the phloem.

Thus in the regions of maximum exchange of water and solutes the elements concerned are very simple, but once the system is acting mainly as a pipeline, as in the lateral and main veins, there is a higher degree of organisation, and the main vein, as we saw in section 4 · 4, may ultimately approximate in structure to the stem.

5 · 2 : 2 Mesophyll and airspace system

The palisade and spongy mesophyll of the typical mesophytic leaf are specialised parenchymatous cells. Green tissues in stems are not so well differentiated, and roots are very rarely green. The manner in which leaf mesophyll develops is responsible for one of its chief characteristics, viz. its airspace system.

In the course of differentiation of the young leaf, mesophyll is formed from layers or sheets of cells parallel to the epidermis. At first epidermis and mesophyll increase in cell number by mitosis, laying down anticlinal walls (i.e. at right angles to the plane of the epidermal surface) and frequently the two or more layers correspond cell for cell at this stage. Later on, mitosis becomes less frequent in the mesophyll layers than in the epidermis, and the mesophyll cells start to increase in size.

When observed in a surface or in a paradermal section the cells of the upper layer (or layers) of palisade mesophyll remain approximately isodiametric, although by now they are elongating in a direction at right angles to the leaf surface, and are beginning to become columnar (i.e. pillar-shaped). Cells of the lower (abaxial) spongy mesophyll layers cease dividing even sooner and, as the epidermal

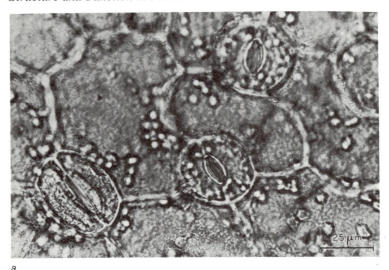

a

b

Fig 5·3 Stomata in a leaf of *Helleborus sp.* (cf. also Fig 5·4).
a. Surface view of the lower epidermis showing three stomata with different apertures. The guard cells contain chloroplasts, but the larger granules in the epidermal cells are not chloroplasts. [Ph.]
b. Surface view of the leaf of *Helleborus sp.* looking through the epidermis into the spongy mesophyll, and showing the relation between the stomata and the air space system (as).

cells and the whole leaf lamina expand, become considerably separated from one another, except in certain areas of contact. In consequence they become honeycombed with airspaces. (See Fig. 4·14d and 5·3b).

By this time the epidermal cells also have ceased mitosis and are themselves expanding, though they are now out of phase with the palisade mesophyll, and it is towards the end of this phase that palisade cells become significantly separated to produce an airspace system which is less extensive than that of the lower mesophyll, but can be well seen in upper paradermal, or better still in epidermal sections where the relationships between the two layers, epidermal and palisade, can be more easily discerned.

Whilst palisade and spongy mesophyll cells differ quite considerably in shape, functionally of course they are identical. All are thin-walled, with fairly prominent nuclei. Their chloroplasts are discoid, of the order of 5 μm in diameter, and having a lamellar ultra-structure; they may contain transitory starch. There are active mitochondria present in most green cells. In the palisade cells the chloroplasts take up a position best suited to the prevailing lighting conditions; it should be observed here that strong light destroys chlorophyll. It is interesting to compare sun and shade leaves in a plant like beech, where, under conditions of higher illumination, the palisade cells are better developed; intercellular spaces are less well developed; the epidermal cells do not extend so much during development as in shade leaves, and therefore there are more stomata per given area; the cuticle is generally more prominent and the rate of transpiration smaller than in corresponding shade leaves from the same plant. Sun and shade plants are discussed again in the section on pigments (see 5·6).

5·2:3 Epidermis and stomata

The epidermis provides a continuous barrier of cells between the mesophyll and the external atmosphere. This barrier is implemented by the development of a cuticle (see also 1·2) and some degree of cutinisation of the outer cellulose wall. The waxy cuticle may be thrown into irregular ridges or folds; it is water repellent and drops of rain stand separately upon it without running together or spreading over the surface. Investigations with the electron microscope on some species (e.g. cabbage) show that the waxy bloom present is due to irregular extrusions of cutin material. All such features make the leaf less wettable, and whilst this may help the plant to avoid the blocking of stomatal pores, it does sometimes present problems to the farmer and gardener, who have to use wetting agents in order that sprays may spread on the leaf surface.

In an ordinary mesophyte with a cuticle of the order of 0·5 μm in thickness, less than 10% of the total water loss may be passing out across the cuticular barrier, and less than 1% of the CO_2 used diffuses in by the same route. The remaining gas exchange is limited to the stomatal pores. In a xerophyte such as the cactus *Opuntia* the cuticular loss may be as little as one two-hundredth of the cuticular loss in a mesophytic leaf such as oak, and only one thousandth of that in *Impatiens noli-me-tangere* (the common balsam) which has a very thin cuticle.

In a vertical section of a leaf such as privet (Fig. 4 · 15) the epidermal cells appear rectangular in section, they fit closely to one another and they make contact with cells of the mesophyll, and indirectly with the cells of the vascular system. Epidermal cells normally lack chloroplasts (except in some aquatic plants) though of course chloroplasts occur routinely in the stomatal guard cells. There is some evidence that epidermal cells may transmit water from the veins to the mesophyll cells in contact with them. Certainly many epidermal cells have prominent pitting in their anticlinal cell walls. Surface views of the epidermis (e.g. Fig. 5 · 3) frequently show the cells to have a sinuous outline, developed in the last stages of expansion, and this must add to the cohesive strength of these tough protective skins.

Stomata: As already stated (p. 69) the epidermal cells continue to divide after mitosis has slowed to a standstill in the mesophyll zone. In many succulents the epidermis may remain meristematic for years. Certain cells of the young epidermis called *stomatal initials* lay down an anticlinal wall (perpendicular to the surface) after a mitosis. Part of the middle lamella of this new wall breaks down to give a pore bounded by two cells, and these cells, the *guard cells*, eventually elongate in varying degree.

The guard cells become differentially thickened in a manner which varies from species to species; the general pattern is that cellulose is laid down on the guard cell walls around the rim of each pore, above and below, whilst a relatively thin-walled region is left immediately adjacent to the pore. In each guard cell a thin-walled region, remote from the pore, acts as a hinge, and when

70

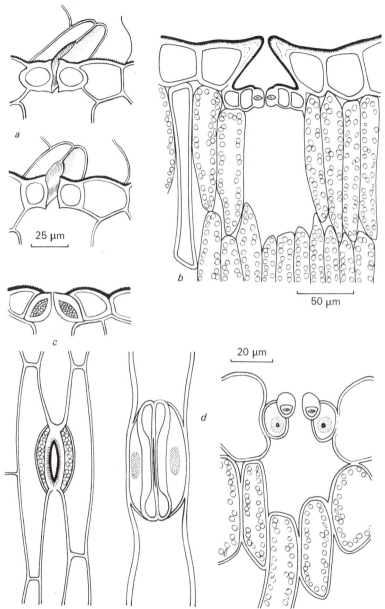

Fig 5·4a. Three-dimensional diagrams of a stoma of *Helleborus sp.*, nearly closed and fully open.

b. T.S. of a stoma of the xerophyte *Hakea sp.* (cf. Fig. 5·1).

c. T.S. and optical view of a stoma of *Iris germanica.*

d. T.S. and optical view of a stoma of *Dactylis glomerata* (cocksfoot grass).

71

turgor is lost in the guard cells, collapse is restricted to a movement of the thin-walled regions adjacent to each pore, which becomes closed where the guard cells meet. This also involves a longitudinal relaxation of the guard cells; the whole stoma becomes longer and narrower as it closes, and so similarly does the pore (Fig. 5 · 3b and 5 · 4a). Other examples of stomata may be seen in Fig. 5 · 4b, c and d.

5 · 3 *Some facts about stomata and diffusion through pores*

Plants with dorsiventral leaves tend to develop the bulk of their stomata on the lower abaxial surface of the leaf. Common exceptions to this may be seen in the leaves of clover, willow and plantain. Grasses and many other monocotyledons develop stomata on both sides of the leaf. Water plants with floating leaves (not unreasonably) often develop stomata only on the upper exposed surfaces.

There is considerable variation between plants of the same and of different species in the number of stomata per cm^2 of leaf surface (*stomatal frequency*). For any given species this will clearly depend on the age of the leaf up to full expansion, since the stomatal initials themselves become more widely separated by the expansion of the epidermal cells.

From the physiological point of view, we need to examine the importance to the plant of its stomatal distribution pattern, and it is useful here to recall experiments of the type first carried out by Brown and Escombe, but since repeated and confirmed many times. If the open ends of tubes containing soda-lime are covered with stout aluminium foil, the foil may be pierced by holes differing (*a*) in total number, (*b*) in diameter and (*c*) in distance apart from one another. It is possible to set up a series of tubes in which a fixed percentage of the area of the foil (the so-called septum) is occupied by any number from one to many perforations. The septum can be pierced in such a way that the holes are close together or widely spaced. Weighing the tubes at intervals gives a record of the rate of entry of CO_2. If water is placed in the tube, one can measure instead the rate of loss of water vapour to the exterior under standard conditions.

Fig. 5 · 5 summarises some of the results that have been obtained by this method. It can be seen that for pores of decreasing diameter the water that is lost is proportional *not* to the cross-sectional

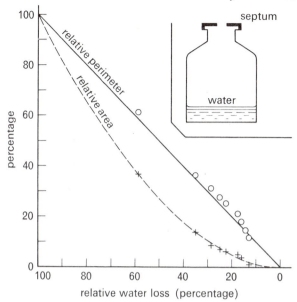

Fig 5·5 Sayre (1926) measured the loss of water across a series of septa, each perforated with a hole of different diameter (see inset for method used). The relative water loss for each size of hole was then compared with (a) the relative area of the pore and (b) the relative length of its perimeter. Each of these parameters is plotted on a percentage scale, and the diagonal represents the condition of best fit. The data confirm that diffusion across a pore is best related to its perimeter rather than to its area.

area of the pores but to their circumference. Since stomatal pores are elliptical in section, the ratio of perimeter to area is even greater than for the artificially made circular perforations.

If the pores are placed more closely together in the septum, it is found that the rate of water loss falls off. In one such experiment, pores that were separated by a distance of five times their diameter lost water at a rate of 60 mg per hour, whilst with pores of the same size separated by twenty diameters only 6 mg per hour of water was lost. The percentage area of the septum occupied by the closely spaced pores fell as the result of wider spacing from 3·38 % to 0·18 %, i.e. by a factor of nearly twenty, whilst water loss diminished to one tenth.

This and other features of diffusion through pores can be shown in diagram form as in Fig. 5·6. With the airspace in the water container at 100% humidity, there is a water vapour gradient outwards to the atmosphere, and watèr molecules pass from regions of high to regions of lower concentration or percentage humidity. It will be noticed that the large pore A gives a wide spread of concentric shells down to 60% humidity; the spread becomes considerably smaller for the smaller pores B and C and the concentric

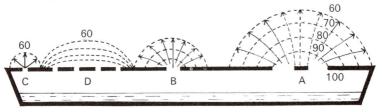

Fig 5·6 A diagram to illustrate some of the features of diffusion through pores which are discussed in the text.

shells are closer together (i.e. the diffusion gradient is steeper). If the centre only of pore A were to be blocked as indicated (see dotted lines) the bulk of the water loss from the perimeter of the pore would hardly be interfered with; water molecules diffusing from the centre of a pore collide with the molecules escaping more peripherally, and those nearest the rim encounter least obstruction. Note that the pore B of the same size as the obstruction in the centre of A, allows a much wider spread of its diffusion shells, provided that there is no interference from adjacent shells. The effect of interference is seen at D, where pores equal in size to pore C are spaced from one another at a distance of less than four diameters. If the pores are spaced more widely than ten diameters apart the diffusion shells do not interfere with one another.

However, we need also to see how such data compare with data for living plants. Most stomata, even when fully opened, have apertures that are smaller than those used in physical investigations. The stomatal frequency and the size of individual stomatal pores may vary even within the same species. In wheat, to take but a single example, the adaxial surface of the leaf may bear between 3000 and 6000 stomata per cm^2, and the abaxial surface between 1200 and 4000. The dimensions of the pore are about 38 μm × 7 μm, so that at full aperture the pore area is just over 200 μm^2, and the pores then occupy between 0·5 and 1% of the leaf surface. Even with the pore area at only 0·5% of the whole leaf area, water loss is unlikely to be less than about 50% of that from a free-water surface of equivalent area to the leaf. The significance of this is that water vapour (*and* CO_2) can traverse the stomatal pores in a given leaf area almost as fast as if the epidermis were not there; yet the stomatal closure mechanism is always available to control excessive loss. The main problem for the leaf is to reconcile the need for the inwards diffusion of CO_2 for photosynthesis with the dangers of excessive diffusion of water vapour in the outward direction.

5 · 4 *Mechanism of stomatal movement*

A good deal has been discovered and written about the mode of action of stomata, but it is doubtful even now whether we know the whole story. Fundamentally guard cell movement as described in 5 · 2 : 3 involves turgor change. A closing stoma has guard cells that are losing water, for whatever reason, whilst an opening stoma is taking in water which, by reason of the differential thickening referred to above, is causing the guard cells to bulge apart. The classical explanation for the changes that are observed states that in the light CO_2 is used in photosynthesis; this increases the pH or decreases the acidity of the cytoplasm of the guard cells, and under these circumstances enzyme action is favoured which hydrolyses starch to substances (such as hexose) of greater osmotic effectiveness; water then moves into the cell osmotically in response to steeper concentration gradients (p. 11) and the cells bulge, thus opening the pore. In darkness, the reverse occurs and under the increasingly acid or lower pH conditions due to accumulating respiratory CO_2, the formation of starch from glucose is favoured and the turgor drops. In plants where starch is not found, presumably the formation of other polysaccharides or oligosaccharides leads to a lowering of turgor.

By the use of labelled CO_2 it has been possible to show that the rather peculiar chloroplasts that are present in the guard cells do actually have weak photosynthetic activity, and that they might therefore account for some of the changes observed. It is of course in any case probable that they function like any normal starch-manufacturing plastid (amyloplast) and are responsible for bringing about any starch–sugar changes that occur. A significant fact is that the stomata open widely when the level of the CO_2 present in the sub-stomatal cavity and the intercellular airspace system drops below 0·01 %. The normal level of atmospheric CO_2 is 0·03 %, and during photosynthesis, i.e. under illumination, the level of CO_2 is kept low. However it still has to be explained how it is that stomata open in the dark when the intercellular space system is flushed with CO_2-free air. Recent work suggests that the opening mechanism is likely to be operated by a potassium pump, working in such a way that the K^+ level rises in the guard cells if ATP is made available. This raises the osmotic pressure of the guard cells and the stomata open.

We have also to reckon with the effects of water deficit in the leaf. Most species respond by stomatal closure when the leaf water content drops below a certain level. In the broad bean, stomata close when the level drops by 3 to 97 % of full saturation, though for many other herbaceous plants water content may drop by 10 % or more before stomatal response is seen. Probably the response to change in water content is the most important factor concerned, and it may even be that the diurnal changes in starch and sugars observed in the guard cells are merely (and more slowly) reflecting and stabilising the faster changes that occur as the result of response to change in leaf water content or in substomatal CO_2 levels. There is little doubt that the guard cells are very sensitive—witness the fact that stomata will respond by a shockwave of closure across a *Pelargonium* leaf when it has been damaged at the margin. Presumably in such a case turgor loss is the result of a sudden change in guard cell permeability, but we simply do not know how this comes about, or even whether it is related to the normal movements observed.

It is worth commenting again at this stage that no plant can grow and succeed without CO_2 for photosynthesis, and it is therefore in the plant's best interests that the stomatal apparatus should remain open and capable of transmitting CO_2 by diffusion until the last moment, when danger by excessive water loss becomes threatening. Even so the daily pattern of opening and closing is quite variable amongst plants. The normal early morning opening rising to a peak round about noon, and falling off again by mid-afternoon or a little later, may in some plants (clovers, etc.) be modified by a midday closure and sometimes a partial opening at night; other plants like potato may keep their stomata open day and night. Since many plants are known to be able to fix CO_2 in the dark, although there is no energy gain as in photosynthesis, it is conceivable that plants which have to keep their stomata open at night but closed by day, put temporarily dark-fixed CO_2 into use by day as part of the photosynthetic process.

5 · 5 *The measurement of photosynthesis*

The basic equation for photosynthesis may be written in the form:

$$6CO_2 + 12H_2O^* \rightarrow C_6H_{12}O_6 + 6H_2O + 6O_2^*$$

which is a convenient way of stating that the oxygen evolved is derived from the water present and not from the CO_2 taken into the system. From the physiological point of view, estimates of photosynthetic rates may be obtained by measuring over a suitable period: (*a*) the quantity of CO_2 that is fixed, (*b*) the quantity of oxygen which is liberated, or (*c*) the increase in dry weight due to the production of hexose and substances derived from it. Although metabolic water is involved, there is clearly too much water already present for this to provide either a convenient or an accurate means of determining photosynthesis.

The uptake of CO_2 can be determined very conveniently by what is called the split-stream method. In this method, air (containing 0.03% of CO_2) is passed over the photosynthesising material at a constant rate. The CO_2 which remains in the gas stream may be absorbed in dilute alkali (generally barium hydroxide), and according to circumstances its absorption may be followed electrometrically (change in conductance of the absorbing solution) or titrimetrically (baryta water is titrated against dilute hydrochloric acid using phenolphthalein as an indicator), or colorimetrically (using an indicator which changes colour as CO_2 lowers the pH). A control airstream passing at the same measured rate can be monitored for CO_2, and the differences observed are a measure of the CO_2 absorbed in photosynthesis. Nowadays it is customary for the split-stream method to be combined with the use of an infra-red gas analyser, which directly monitors the CO_2 present in terms of its capacity for absorbing energy of long wavelength (i.e. infra-red).

Most of the earlier research on the gas exchanges in photosynthesis were carried out using what are known as manometric methods. An illuminated tissue or algal culture gives off O_2 and consumes CO_2. In a closed system this can be observed as a change in manometric pressure in a system kept at constant volume, or alternatively as the change in volume of a system kept at constant pressure. The most usual equipment is the Warburg apparatus the use of which demands that all precautions are taken to account for fluctuations in temperature and barometric pressure. Manometric methods make it possible to measure both CO_2 uptake and oxygen output, but the latter can also be monitored in a variety of ways, most of them beyond the scope of this book.

In class experiments using water plants, whilst a bubble-counting technique is suitable for comparing, say, the relative rates of photosynthesis under different light intensities, it is desirable that some analysis should be undertaken of the composition of the gas evolved, e.g. in a graduated capillary tube using alkaline pyrogallol to absorb the oxygen. The air in a closed system containing an illuminated terrestrial plant can similarly be analysed for change in the percentage composition of CO_2, O_2, and nitrogen, and a good deal of useful information may be obtained in this way.

Finally it is possible to assess photosynthetic yield in terms of dry weight increase. For this purpose it is best to sample from a large population of plants (e.g. the leaves from a privet hedge) and to use carefully matched pairs of leaves; the first of a pair is taken at the beginning of the day and the second at the end. Given a large enough sample, it is a fairly straightforward matter to demonstrate an increase in dry weight at the end of a photosynthetic period. This kind of measurement, provided that the matching is done with care, is independent of the variables such as water content, nitrogen content and so on, which can so often complicate the picture in other ways.

In making measurements of this kind we must be prepared to accept that they only define *apparent* photosynthetic rates, since all tissues are simultaneously undergoing respiration. The real photosynthetic rate would measure all of the oxygen evolved, whereas in fact some oxygen is used in respiration to derive energy from hexose present. It will be clear that most of the methodology described is equally applicable to the measurement of gas exchange due to respiration alone.

5 · 6 *The pigments of photosynthesis*

Many aspects of cell metabolism contribute to the efficiency of the photosynthetic process; together they comprise the internal factors, of which only one aspect is going to be considered here, namely the pigments of the chloroplasts. Thanks to the electron microscope we now know that the chloroplast is a lamellate structure, in which membrane-bound cavities contain the pigment system. The cavities are concentrated into disc-shaped stacks known as *grana*, each about 0·6 μm in diameter which are embedded in a *stroma* or matrix. At an even higher level of resolution, it can be demonstrated that inner surfaces of the grana lamellae are lined with flattened spheroidal bodies called *quantasomes*, about 20 nm in diameter and 10 nm high.* It is believed that these bodies represent the smallest units within which the whole gamut of the light reactions can take place. In general it is believed that the *light* reactions of photosynthesis are restricted to the grana, whereas the *dark* reactions mostly occur in the stroma.

In looking for links between structure and function, we need to know more and more about the spatial relationship between molecules. In no field is this more important than in the field of photosynthesis, where the impact of quanta of light upon a critically spaced system of pigments and their supporting molecules enables radiant energy to be transferred and trapped with the aid of enzyme molecules in the formation of new chemical bonds.

In the development of chlorophyll within a chloroplast exposure to light is necessary. Even in light the plastids do not normally start to become green until they have attained about half of their final size. Most higher plants and the green algae show the development of chlorophylls a and b, together with a variety of accessory fat-soluble carotenoid pigments, i.e. carotenes and xanthophylls, which absorb light in the blue end of the spectrum, and play some role in making this energy available for part of the photosynthetic process.

The synthesis of chlorophyll is under genetic control but depends also upon nutritional factors. Apart from the magnesium and nitrogen that are essential components of the chlorophyll molecule, lack of iron, and also of the micronutrients manganese, copper and zinc, can be a cause of chlorosis or chlorophyll deficiency. There are many instances known of genes which as double recessives give rise to albino strains (note for example the albino forms of corn, *Zea mais*). Such genes are clearly lethal, and the seedlings can only grow whilst there are reserves left in the seed.

* 1 mm = 10^3 μm = 10^6 nm = 10^7 Ångstrom units.

Structure and Function in Plants

Light-starved seedlings are described as etiolated and are yellowish in colour. They will only develop green chlorophyll if light is allowed to convert the precursor protochlorophyll that is present in the etiolated leaves. This reaction is triggered off by the near-red far-red phytochrome mechanism (see p. 199 et seq).

On the other hand, excessive light inactivates chlorophyll, and presumably the movement of chloroplasts to the more shaded parts of a highly illuminated palisade cell does something to offset this situation. Some plants, e.g. the rubber plant (*Ficus elastica*),

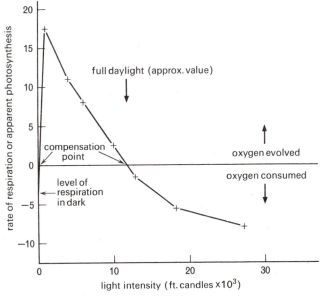

Fig. 5·7 The effect of increasing light intensity on apparent photosynthesis in the green alga *Chlorella sp.* Progressive breakdown of the chlorophyll occurs with increasing light intensity; *Chlorella* is really a shade plant.

develop a multiple leaf epidermis which acts as an extra water-storing shield and may have a partly protective function. Figure 5·7 shows how the relative rate of photosynthesis in the alga *Chlorella* falls off with increasing light intensity up to a level equivalent to about three times the intensity of full sunlight. The fall-off reflects the progressive photo-inactivation of the chlorophyll. Light in-activation seems to be diminished in leaves with a high sugar content, and there is little doubt also that most organisms, whether in the sea or on land, adapt their pigment 'equipment' to their situation. Shade leaves not only contain less chlorophyll, but they are relatively more efficient when working at low light intensities. Sun leaves by contrast are relatively less efficient when the light intensity is at a low level, even though they may possess a higher content of chlorophyll.

5 · 7 *External factors controlling photosynthesis*

5 · 7 : 1 Light

Some of the effects of light in promoting both the initial synthesis of chlorophyll, as well as its inactivation, have already been considered. Basically however, we should concern ourselves with the effects of light in terms of its energy content, and the efficiency with which that energy can be trapped and converted to chemical bond energy. Light varies in intensity and in quality (i.e. spectral composition), and it should be understood that short wavelength blue and ultra-violet light has a higher energy yield than long wavelength red and infra-red. The intensity of light can be measured in a variety of ways. The normal selenium 'barrier-layer' cell (as used in photographic exposure meters) measures radiant energy in terms of micro-amperes and has a range of sensitivity not much greater than that of the human eye. In other words it does not respond to light in the infra-red and ultra-violet sectors of the spectrum. In this sense it is quite well suited to measurements of light intensity in experiments on photosynthesis, for the green plant, too, makes little use of light of these wave-lengths. Such cells are calibrated with reference to a standard light source, and the intensity is expressed in lumens m^{-2}. However, for most physiological purposes it is more appropriate to use, as a measure of radiation, SI units of energy such as $J\,cm^{-2}\,s^{-1}$, since radiant energy of all wave-lengths may be converted into heat when absorbed by a dark surface, and this is of obvious importance, for example in transpiration measurements.

In this country, the illumination of a surface by direct midsummer sunlight may correspond to an incident energy level of about $0.036J\,cm^{-2}\,s^{-1}$, and this value is bound to vary with latitude and with season. For a given locality at a given season the diffuse illumination from a clouded sky may be approximately half that of direct sunlight. Clearly yet more energy will be absorbed when light passes through a leaf canopy. For example, a stand of evenly spaced pine trees may cut indirect illumination to one fifth of the value for diffuse light as measured above the canopy, whilst under

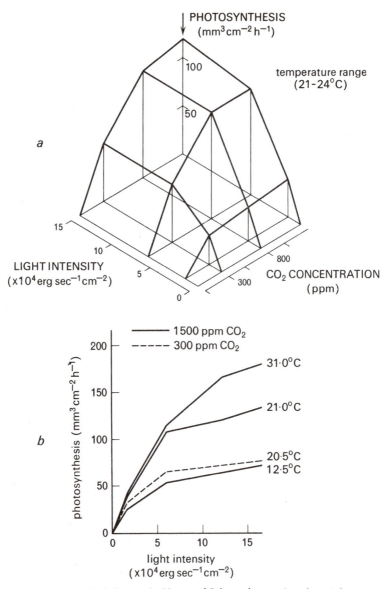

Fig 5·8 Photosynthesis by attached leaves of *Solanum lycopersicum* (tomato).
a. A three-dimensional diagram showing the interaction between light and CO_2 concentration over a restricted temperature range (between 21° and 24°).
b. A graph showing the variation in photosynthetic rate with light intensity at three temperatures, in the presence of 1 500 ppm of CO_2. Note that at 300 ppm of CO_2 (i.e. atmospheric level) and at 20·5°C, the rate levels off in spite of increasing the light intensity.

82

a canopy of beech trees, radiation is reduced to about 5% of the diffuse light outside. It will be seen how this may affect the growth of plants on the woodland floor. Furthermore, the spectral quality of the light is changed as it passes through the green canopy. In a similar way, during the penetration of light into the sea, the shorter wavelengths of light are absorbed more and light in the blue-green part of the spectrum penetrates further. At a depth of five metres the energy level may be reduced to one tenth of that at the surface, depending on such factors as turbulence and turbidity of the water.

If no other factors are limiting, the rate of photosynthesis increases with increasing light intensity up to a point at which the cells are said to be light-saturated. Beyond this point, as already seen, photo-inactivation becomes important. If another factor is limiting (as shown for CO_2 in Fig. 5·8) the rate curve flattens off sooner. The rate of a given process is not necessarily determined by all of the essential factors involved, but may be limited by any *one* of these that is deficient (as well as by those such as light which in excess may be retarding). Blackman first drew attention to the principle of limiting factors at the beginning of this century.

The quality of the light received by the plant is also important, and a number of investigations have been made of the relative photosynthetic efficiency of light of different wavelengths. Figure 5·9 shows the photosynthetic response of wheat plants to light in the visible spectrum, plotted in relation to the absorption characteristics of chlorophylls a and b, and carotene. In observations like these it is most important that the plant should receive equal amounts of radiant energy for each selected wave-band, and also that light should be the only limiting factor. It should be obvious that the quality of light received by a plant will vary with latitude, and season (not to mention with cloud, dust or smog). The higher the latitude the greater the loss of light from the blue end of the spectrum. When the sun is furthest away in winter, its light is appreciably more red. The fine red sunsets associated with the polluted atmosphere of some of our industrial cities pay their own testimony to the effects of dust in the atmosphere. The green light which filters through a leaf canopy is less efficient than red or blue light, but nevertheless makes some photosynthesis possible, and on the woodland floor this and the transient flecks of penetrating sunlight may enable a ground flora to be maintained.

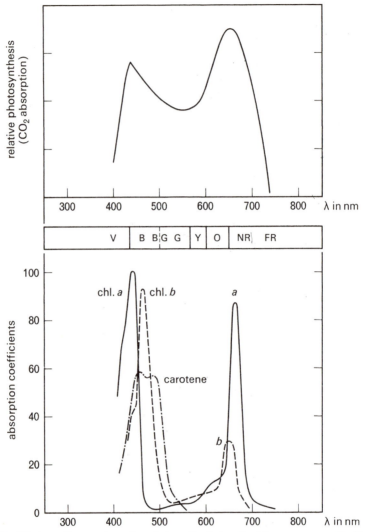

Fig 5·9 An action spectrum for photosynthesis in wheat leaves, compared with absorption spectra for chlorophyll *a*, chlorophyll *b*, and carotene. The peaks coincide in the blue ($\lambda = 450$ nm) and the near-red ($\lambda = 680$ nm) parts of the spectrum.

5·7:2 Carbon dioxide

The atmosphere contains 300 parts per million of CO_2 (or 0·03 %), although higher levels may be detected under standing crops and in forest litter. There is some reason to believe that, as a result of the evolution of the photosynthetic process and of its successful adaptation to terrestrial conditions, an earlier much higher level of CO_2 in the atmosphere has since been reduced. By contrast the oxygen level now stands at about 20·9 %. Most scientists accept that in the earliest stages of the evolution of life on earth oxygen was probably absent.

Photosynthesis thus plays a central role in the cyclic turnover of carbon compounds; the carbon that is culled from the air is incorporated into succeeding members of chains of inter-dependent organisms. As each individual member of a food chain respires, so it converts energy to its own use, and loses carbon dioxide back into the atmosphere. Primary producers trap energy and fix carbon; consumers use some energy and pass on some energy, and some of the carbon is returned into the atmosphere as CO_2 at each stage. It has been estimated that the fixation of CO_2 may exceed 150000 million tons per year, and only one tenth of this is fixed by the more efficient terrestrial plants; much of the fixation takes place in the sea with the help of phytoplankton, and it is to the sea that man must look more closely if he wants to make better use of nature's resources for food.

Carbon dioxide concentration is most frequently a limiting factor in photosynthesis, and most plants can easily handle higher-than-atmospheric concentrations. The yield of some greenhouse crops (such as tomatoes) may be improved by boosting the level of CO_2 available to the plants. The forbidding economics of the situation involve the cost of making glasshouses rather more gas-proof than they normally are, and this implies a better regulation of temperature and humidity. Most plants show increased utilisation of CO_2 if air is moved over them rather than allowed to remain stagnant. At the same time it should be noted that an excess of CO_2 may prove to be limiting, partly by its effects in promoting the closure of stomata, and partly through interference with the general metabolism of the cells containing chloroplasts.

5·7:3 Water
It is an obvious fact that water is required during photosynthesis, and one might well imagine that the amount required could only be but a small fraction of the total turnover. However, the rate of photosynthesis falls off for two reasons as the leaf passes into a state of water strain. The first reason is fairly obvious, namely that with water loss the stomata close, and little or no CO_2 is then

available for the process. It has already been seen how plants differ in their stomatal response to water loss (p. 75). The second reason is that the photosynthesising cells themselves need to be fully turgid; it is reported for example that in the presence of an osmotically active substance such as mannitol, photosynthesis is reduced at or about the point of limiting plasmolysis in the leaves of the pondweed *Elodea*, even though more than adequate CO_2 is available in solution. There are many records available which indicate that trees and crops fall off in photosynthetic activity as their water supply becomes impaired.

5 · 8 *Compensation point*

It has already been pointed out that the activity of chloroplasts can sometimes be adapted to higher or lower light intensities. However, as a general rule shade plants do best in low light intensities, just as by contrast sun plants will not easily survive the light competition of taller dominant neighbours. The metabolic activity of a leaf exactly balances out when the respiratory output of CO_2 is just used up in photosynthesis. The term *compensation point* is used to describe this situation, and it should be obvious that since CO_2 is nearly always in short supply, light and temperature are the factors that primarily determine the compensation point.

Thus the compensation point of a plant at a given temperature (which determines the rates of the dark reactions of photosynthesis as well as the rates of respiratory processes) can be expressed in terms of light intensity. It is clear that the lower the light intensity, the lower must be the temperature for balance to occur. Some of the brown seaweeds demonstrate this point admirably, for in culture they will not grow under favourable light intensities until the temperature is lowered to about 5°C. Growth, as must seem obvious, cannot take place until photosynthesis is in excess of respiration.

A shade plant such as wood sorrel (*Oxalis acetosella*) is adapted to shade conditions (i.e. it photosynthesises most efficiently under light of lower intensities), but it also has the advantage of extra CO_2 provided by the decomposition of the woodland litter in which it is growing. It is tolerant of these conditions, just as beech seedlings are tolerant of dense parental shade, and can establish themselves under conditions in which the light intensity under the leaf canopy can be as little as 1–2% of full daylight outside.

6 Transpiration and water movement in the plant

6 · 1 *Introduction and anatomical considerations*

It was emphasised in the last chapter that some of the characteristics of leaf structure which most influence photosynthesis are of similar importance in regulating the process of water loss. The control of gaseous diffusion is no less relevant for water vapour efflux, or for the movement of oxygen, than it is for carbon dioxide influx.

LEAF STRUCTURE. Little need be added to what has already been discussed in previous sections. Water loss from the leaf, as determined by the drying power of the atmosphere, is chiefly modified by two anatomical factors: (*a*) the resistance to water flow offered by the cuticle and (*b*) the efficiency of the stomatal mechanism (see Chapter 5). The inclusion of stomata in the epidermis (which may be seen in organisms as far down the evolutionary scale as the mosses) makes control possible at a most effective locus, viz. between the more or less saturated intercellular airspace system beneath the stomata, and the ambient atmosphere which lies beyond the diffusion shells around the stomatal pores, merging with the boundary layer of air adherent to the leaf surface.

MOVEMENT IN THE STEM. The resistance to water movement through the stem varies somewhat with stem anatomy, being greatest in coniferous stems where water transport is restricted to non-perforate tracheids; the increased resistance to flow offered by these elements can be ascribed to the fact that water has to move from one tracheid to another mostly across the thin cellulose membranes of the connecting bordered pits. In Fig. 6 · 1a–d differences are shown between simple and bordered pits, but in both instances the middle lamella bridges the pit cavity and offers a lower resistance to water movement than does the cell wall. Trees such as oak have ring-porous wood, i.e. most of the vessels present are

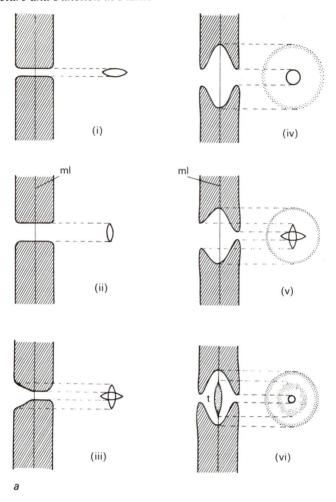

a

Fig 6·1 Pits. (cf also Figs 3·1, 3·2 and 3·3). See opposite page.
a. Diagram illustrating simple and bordered pits and their appearance in optical section and in transverse section.
   (*i*) and (*ii*) show how simple pits may appear to vary in width according to their orientation (ml middle lamella). (*iii*) shows a simple pit with crossed slits. (*iv*) is a bordered pit, (*v*) a bordered pit with crossed slit apertures, and (*vi*) is a bordered pit with torus (t).
b. A sectional view of simple pitting in heavily thickened walls of the ray parenchyma of *Liriodendron sp.*
c. A sectional view of bordered pitting in a vessel of *Liriodendron sp.*
d. An electron micrograph of a pit field from a young, primary cell wall of a coleoptile of *Zea mais*; note that it is crossed by some primary cellulose microfibrils.

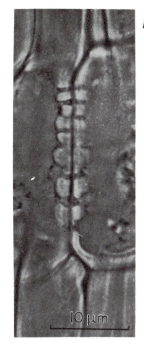

b

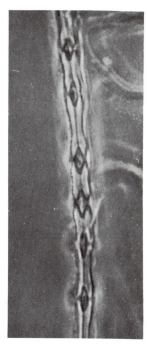

c

10 µm

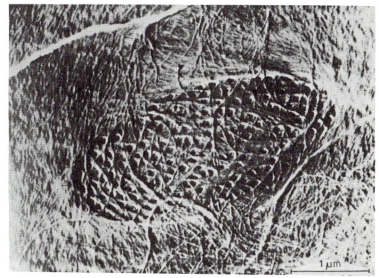

1 µm

d

associated with the spring wood. The uninterrupted length of vessels available for the passage of water is generally greater in ring-porous woods than in diffuse-porous woods, where, as the name suggests, vessels continue to be added throughout the season. For example, individual vessels in oak may run without interruption for much of the length of the trunk. This can be demonstrated by the injection of Indian ink into the lumen of vessels whose contents are under tension. With the entry of fluid at atmospheric pressure the tension is released, and the injected particles may subsequently be traced over 6–7 metres of uninterrupted vessel. Diffuse porous woods rarely show such long continuous pathways.

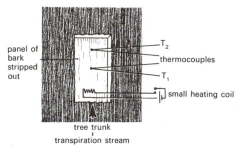

Fig 6·2 Diagram illustrating a method of investigating the rate of flow of the transpiration stream in a tree.

Confirmatory evidence comes from the application of an ingenious experimental technique which does not involve damage to the vessels themselves, as in injection experiments. This technique involves the use of sensitive thermo-couples (T_1 and T_2 in Fig. 6·2) in conjunction with a small heating coil (C). The cambium and wood are exposed by the careful removal of a section of bark, and the coil and thermo-couples are arranged one above another as shown, with appropriate heat insulation. The small rise in temperature conveyed to the ascending water stream is detected first at T_1 and subsequently at T_2, thus making a comparison possible between rates of flow in different trees. The range of flow rates found may be briefly summarised:

Ring-porous woods from 2000 to 4500 $cm h^{-1}$

Diffuse-porous woods from 200 to 500 $cm h^{-1}$

Coniferous woods about 60 $cm h^{-1}$

Whilst flow rates must in part be determined by the mean cross-sectional area and the length of the vessels involved, this type of experiment also lends support to the view that the bulk of the sap is moved in the peripheral vessels nearest to the cambium. This would help to explain how the capacity for water conduction remains adequate to its task, even when the more central tissues (especially the heart wood) are blocked with cellular ingrowths from the living ray-cells, with mineral deposits and gums, and by the comparatively frequent development of gas bubbles.

6 · 2 *Some physico-chemical considerations*

Loss of water by evaporation from the leaves results in a decrease in the chemical activity or 'effective water concentration' of the mesophyll cells (see also p. 103) and thus a steepening of the overall water concentration gradient between mesophyll cells and soil water. This gradient is transmitted osmotically from the mesophyll cells to the water in the dead elements of the xylem, which is thereby placed under tension. In its turn this state of tension will bring about movement of water from the living cells of the root; we shall see in Chapter 7 how this in its turn may be transmitted to the soil water. The 'effective water concentration' of the dilute soil solution is, of course, high, and progressively upwards, from roots via xylem to leaves, there is a lowering of the water activity, resulting as always in flow down the gradient from regions of high to lower activity, i.e. from soil to atmosphere.

It should perhaps be added for completeness here that the *driving force* which results in water movement from one point to another is dependent on the difference between the *partial molar free energies* of the water at each point. This difference, whilst based essentially upon what we have called above the effective water concentration, may be expressed in negative atmospheres (or bars), having units of energy per volume (e.g. joules cm^{-3}). It is a term that is conveniently applicable to all phases of the water system, from soil, via plant, to gaseous atmosphere. It expresses what has been called the *water potential* as the difference in partial molar free energy between water at any given point on the pathway and pure water. If the water potential of a given cell sap has a high negative value, the effective water concentration difference between the cell sap and pure water is similarly high, due to the presence of solutes in the sap. The water potential of xylem sap similarly decreases (i.e. has a higher negative value) the more it is subjected to tension, just as it would increase (i.e. have a lower negative value) if it were placed under compression.

The reason for introducing this concept here is that it has become increasingly necessary for plant physiologists to come into line with physical chemists, so that both may use the same terminology in describing the phenomena of water relations in living as well as in physical systems.

The work done in moving water through the plant makes use of radiant solar energy. This warms the atmosphere and enables it to hold a higher concentration of water vapour; its drying power increases as its relative humidity decreases. Radiant energy absorbed by the leaf, together with the comparatively negligible amount of heat energy dissipated during metabolism, raises the leaf temperature and assists in the evaporative process. The energy level of water molecules is raised to the point at which they can escape from the liquid to the gas phase. Resistance to gaseous flow by diffusion has to be overcome as the water molecules pass outwards down a concentration gradient to the outside atmosphere, whose water vapour content (effective concentration) determines the steepness of the gradient. At all levels the water potential (for liquid or gas) can be expressed in negative atmospheres. The drier the air the steeper the gradient, and the greater the tendency for water molecules to pass outwards from the plant to regions of water potential with higher negative values (lower water concentration). With fully saturated air the gradient is very much reduced, and no flow of water molecules can occur unless the plant supplies energy. Water droplets may sometimes be seen at the tips of plants (e.g. young grass leaves) when transpiration is checked in this way; this is known as water of guttation. Work must be done by the plant in order to secrete water to the exterior under such conditions.

The water potential of the ambient atmosphere above the plant will obviously have a high negative value since in all gases the molecules are widely dispersed, and are not coherent as in the aqueous phase; hence the effective concentration of water here is low, though it is bound at all times to be dependent upon changing temperature and humidity.

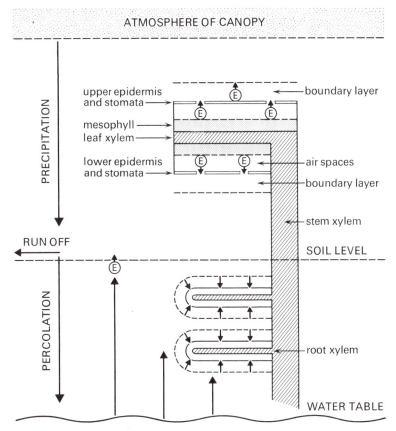

Fig 6·3 A formal diagram to illustrate the soil-plant-atmosphere complex discussed in the text. At points marked Ⓔ evaporation of water requires the provision of energy.

6 · 3 *The soil–plant–atmosphere complex*

It may be helpful at this stage to try to establish a picture of the whole system for plants that are taking in water by root action from the soil and transferring it to the atmosphere via their leaves or other transpiring surfaces. The system may be summarised diagrammatically as in Fig. 6 · 3.

Water movement may take place by bulk (or mass) flow in the liquid phase, or it may involve the slower movement of diffusion in either liquid or gas phase. Wherever there is a change of phase from liquid to vapour in the soil, or at the surface of the leaf meso-phyll, or at a cuticular surface, latent heat of evaporation must be transferred to the system. Below ground the heat capacity of the soil and its contained water is such that there are no large fluctuations in temperature, and the energy for evaporation is made good from the soil system itself. The distribution of heat within the soil system is dependent (*a*) on the absorption of radiant solar energy by the surface layers, (*b*) the slow conduction of thermal energy downwards, (*c*) a very limited amount of convectional movement in the soil atmosphere, and (*d*) the heat energy released by micro-organisms during the decay of organic materials. Most of the liquid-vapour transition takes place at the sun-warmed surface.

In the plant, leaves absorb solar energy and may warm up to an appreciable degree above their surroundings. The energy contribution from the metabolic processes of the leaf itself is for practical purposes negligible. Whatever energy is available accounts for the evaporation of water at or in the mesophyll cell surfaces. It will be remembered that these surfaces are not easily wetted, and it may very well be that evaporation must take place mostly from within micro-capillary pores ending in the cell wall surface. From this point onwards we are concerned with the gaseous diffusional flow of water vapour molecules, though this is still taking place down a gradient of decreasing effective water concentration. Thus the water potential of air with a relative humidity of 50% is something under -1000 bars, which is nearly 20 times the highest values that are normally recorded for leaf mesophyll cells at the top of very tall trees.

The following points may be made with the help of the diagram. Soil water is made available by precipitation, and the soil water which penetrates the soil (some water runs off into drainage channels), does so under gravity; it may eventually add to the existing fluctuating water table. The roots make what use they can of gravity water whilst it is available; after this they can make decreasing use of the water within a prescribed zone around them. In a zone of limited height above the water table (depending on the structure of soil) more capillary water is available, but a plant can only increase its water absorption capacity by growth, that is by extending its root system to make use of the full soil volume that is available to it (see also 7·1). Water passes across the cortical cells of the root either in their cell walls or through their cytoplasm. The route

94

that is taken is probably restricted to the cytoplasm during the passage of water across the endodermis. Water enters the dead cells of the xylem from the surrounding living parenchyma. It can ascend freely in the xylem vessels though it may have to traverse lignified cellulose walls, or the cellulose membranes of pits (p. 87). It may pass out in the leaf across living mesophyll cytoplasm, but could probably travel more easily in the cell walls of the mesophyll. It evaporates at or in the mesophyll surface, and water vapour passes into the substomatal airspaces; water that passes out as a cuticular component of transpiration must evaporate at the cuticular surface. The stomatal pores may help to control the exit of water vapour across the pore to the external atmosphere, and the water vapour must diffuse across the boundary layer of air and water vapour that adheres to the surface of the leaf. The effective depth of the boundary layer depends on the extent of air movement outside the leaf. It is less deep when air movement is greater, and then offers less resistance to diffusional flow. Any plant growing in a community loses its water vapour into an atmosphere which is conditioned by the extent of the plant canopy, its height above ground and its density. Above the canopy, the ambient atmosphere varies with weather conditions, being basically determined by temperature and humidity.

It is generally assumed that the stomata, strategically placed in the first part of the gas phase, constitute the most important mechanism for controlling water loss. Whilst this is probably broadly true, a number of instances have been reported where the stomata stay open, and may indeed on occasion open more widely at the onset of drier conditions. A good example of this kind of 'anomalous' behaviour is provided by sweet corn (*Zea mais*) where the rate of transpiration falls as the evaporation rate rises, but the stomata do not close until the rate of transpiration is quite low. This suggests that other factors may be responsible for reducing water loss, and one suggestion is that the resistance to flow in the mesophyll cells may increase with water loss. If this is so, then somehow the water potential of some part of the system, possibly in the micro-capillaries of the mesophyll cell wall, must rise almost to the level of the water potential in the outside atmosphere.

However, the overall rate of water flow through the plant is first and foremost dependent upon the availability of water in the soil. Soil water tension increases dramatically, i.e. the water potential develops a higher negative value, once the roots have used up the gravity and capillary water available to them. From this point water may become practically unavailable to the plant, which then begins to wilt permanently. The *permanent wilting percentage* of a soil is defined as the percentage water content at which plants fail to recover at night, or if covered to cut down transpiration by day. Under such conditions, growth (especially of roots) is no longer possible, though some very small uptake of water may still occur. Although at one time thought to be approximately constant for most soils, it is now recognised that the permanent wilting percentage varies with the water potential of the leaf sap of the plant under test, which for a wide range of crop plants may show variation between -10 and -20 bars. At the mean value of -15 bars, the water content of a sandy loam may have dropped to 3%, whilst a clay soil may still contain as much as 10% of water; this is a good indication of the capacity of clay soils for binding water which is not available to the plant.

The aerial parts of the plant reduce water vapour loss by cuticular development and by stomatal action; by developing resistance to flow in the aqueous phase (i.e. the symplast), it is possible that they may add another line of defence. In any case at this stage survival is all that matters, and drought resistance depends in the ultimate analysis on the resistance of protoplasm itself to desiccation. Few plants have leaves that can endure more than 20–25% water loss without injury, although resistance to drying may vary with season, and also with the age of the leaf. It is said that the leaves of the olive (*Olea europea*) may lose up to 30% of water before permanent injury sets in, but this is atypical and most leaves are dead long before this stage is reached. Resistance to desiccation is seen at its best in some of the small-celled liverworts and mosses characteristic of dry habitats, and it is well known that herbarium specimens of these that have been dry for years, can be soaked back to life again and resume growth.

7 Water and solute uptake

7 · 1 *Introductory*

In the higher plant the bulk of the uptake of water and dissolved solutes is accomplished by means of a root system. Non-vascular plants such as the mosses and liverworts are dependent upon a capillary water supply; many of the mosses can endure drought because their cells are capable of resisting extremes of desiccation and they recover as soon as rain is again available; yet others of this group, especially the liverworts, will only grow in damp habitats. In a sense the root system of the higher plants is the sole remaining link to an aquatic as opposed to a terrestrial environment. If it is to be accepted that at some time there has occurred a gradual emergence from water to land, with all of its attendant problems of support and water loss, then it is clear that the root system is the only part of the whole system that has maintained its contact with an aquatic environment, viz. the soil water.

The root system has to be thought of in dynamic terms of its capacity for growth. It absorbs water chiefly through those parts of its surface where there is no protective suberisation. If the whole plant system is to absorb enough water it must continuously extend its absorbing surfaces by growth to match the increasing demands of its aerial parts. It must grow in order to exploit the whole soil volume below those aerial parts, by ' threading and rethreading ' the soil spaces in a continuing pursuit of available water. During this process, parts of the root system inevitably come into competition with the roots of other plants. Above ground competition is worked out in terms of the availability of light, but below ground the secret ceaseless battle is for water and the electrolytes that are dissolved in it. Supply depends upon growth vigour, and in its turn this depends upon above-ground metabolism successfully carried out with the water received, and thus food materials are made available for translocation to the growing root apices. This is a highly integrated activity, and the one part of the plant cannot survive without the other. It is an area of plant physiology that is recognised in the concept of the *root* : *shoot* ratio, but in fact we know all too little of this vital and intimate relationship.

7 · 2 *Structure and function in the root*

7 · 2 : 1 The absorptive regions

Figure 7 · 1 shows that in the typical root a phase of active mitosis (i.e. increase in cell number) at the apex is followed by an extension phase (i.e. increase in cell length). As extension growth starts to slow up, so in this region cells differentiate further. Phloem cells appear in a position between adjacent but not yet mature protoxylem groups, and above this level, protoxylem begins to show its character-istic forms of thickening (annular or helical). This type of addition of secondary and subsequently lignified cellulose is eminently suitable for cells whose walls are still being stretched, yet which may be subjected to quite severe conditions of water strain (as water losses are sustained in the aerial shoots above). If the contents of the protoxylem are under tension, and yet the surrounding parenchyma cells are turgid, it follows that protoxylem cells may sometimes be subjected to pressure from the surrounding cells. If the thin primary cellulose walls were not reinforced by rings or helices of secondary wall material (which quickly becomes strength-ened by impregnation with lignin) the protoxylem cells might collapse under these conditions. The later differentiating meta-xylem cells do not have to contend with the last stages of extension growth; they are in fact differentiated to maturity under much more stable conditions, and their patterns of secondary wall formation are different and more varied, and include the formation of bordered pits, and more complex patterns of secondary wall thickening.

The whole of the surface of the young root remains capable of absorbing water at least until suberisation of the exodermis begins and maybe later. The youngest part of the root, with its protective lubricating cap, can absorb water just as readily as the piliferous layer behind. It should be remembered, too, that it is in this absorp-tive region that differential ion uptake occurs; the mechanism for ion selection resides primarily in this very active young region. The piliferous zone can be shown by measurement and calculation to increase the absorptive area of the root surface by between three and five times (and sometimes more); the root hairs (Fig. 7 · 2) make possible the most intimate contact with the minute soil

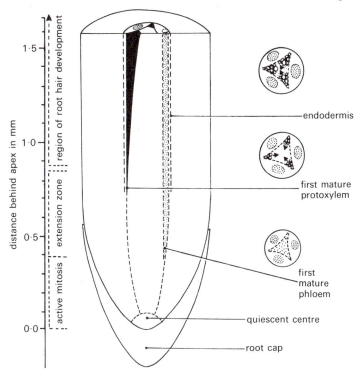

Fig 7·1 Diagram of a root apex showing the approximate relative dimensions of meristematic, extension, and piliferous zones. The first appearance of mature phloem can be seen, nearer to the apex than the first-formed xylem. The differentiation of both xylem and phloem in the root is inwards (i.e. centripetal).

particles and the film of water with which these are associated. By this intimate contact they assist also with the anchorage provided by the whole root system. In the transplantation of plants, care must be taken to move plants with the minimum disturbance of the fine root system, and in addition to guard against excessive transpiration until such time as an adequate absorptive system has redeveloped.

7 · 2 : 2 The adult root

In its anatomy, already described in 4 · 2 : 5, the primary root may be seen to match up to its dynamic requirements by the provision of an anchoring as well as a transport system. The root lies in soil

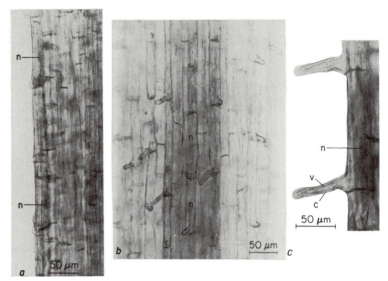

Fig 7·2 Photographs showing the production of root hairs in *Polygonum persicaria*.
a. Papillae arise at the distal end of the piliferous cells.
b. Surface view of a root at a slightly later stage. The darker middle zone is provided by a stained central stele. Nuclei (n) remain in the body of the piliferous cells.
c. The hairs are now obviously vacuolated (v), and particles can be seen streaming in the cytoplasm (c) of the living cell.
In all cases the root tip lies below the lower margin of the picture.

material that is much more dense than itself; it thus has no problems of support as such, although near to the surface of the soil a root must be able to withstand forces of compression *and* tension (see Fig. 7 · 3). It is well recognised that strains of this type, which do not normally involve bending, are best met by the provision of strengthening material to give a central rod-like structure.

From the functional point of view, once the root has differentiated to the point that its root hairs have withered and are replaced by the development of the underlying corky exodermis, it is primarily concerned with an upwards movement of water and a downwards translocation of food materials to sustain growth at the root apices. The differentiating protoxylem elements are strategically well placed next to the endodermis, and their differentiation is contemporary with the nearby root hair region; only later are they reinforced by the addition of larger metaxylem elements.

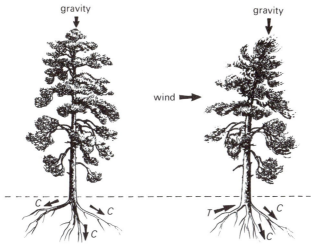

Fig 7·3 A diagram to illustrate the distribution of forces of compression and tension in a tree root system.

By the time that this 'pipeline' system really needs to carry a much larger volume of water, it is further reinforced by the addition of secondary wood (together with small amounts of secondary phloem) by cambial action. The stimulation of cambial activity in woody roots is part of the total wave of cambial stimulation that originates in the developing active meristems of the opening buds and spreads downwards, causing the cambial cells to start into active cell division that eventually reaches right down into the root system.

7 · 3 *The endodermis*

At the boundary between cortex and vascular stele lies the endodermis, and it may well prove that this is one of the most important tissues in the developing root. Because so much important theory is associated with it (though this theory is by no means fully substantiated), it will be described here in some detail. According to its stage of differentiation, endodermis is described as primary, secondary or tertiary. In the primary stage of development, it may be recognised by the development of the Casparian band named after Caspari who described it. This is impregnated with corky material which is deposited within the radial walls (Fig. 7 · 4a). The

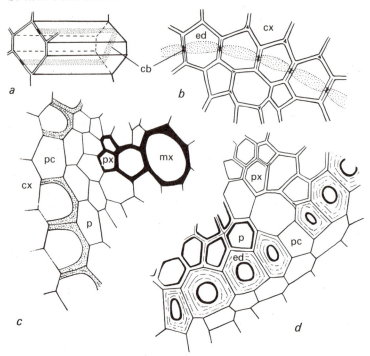

Fig 7·4 A diagram and drawings of endodermis at various stages.
a. An endodermal cell in the primary condition with Casparian bands in the radial walls.
b. Primary endodermis in a young etiolated and plasmolysed stem of *Vicia faba* (cf. the stem endodermis in Fig 4·4*a*.)
c. Tertiary endodermis in the root of *Iris sp.*
d. Tertiary endodermis in an aerial root of *Epidendron sp.* (orchid).

effect of this band is presumably to limit the radial movement of water and solutes to the living cytoplasm of the endodermal cells; water cannot pass radially inwards or outwards along the cellulose walls if they are 'waterproofed'. The cytoplasm is firmly attached to the radial walls at this point (as can readily be seen by plasmolysing a section of a young primary root, see Fig. 7 · 4b) and electron microscopy demonstrates that the cytoplasmic contents of adjacent cells of the endodermis show lateral continuity with one another via minute pores in the radial walls which are traversed by cytoplasmic strands (*or plasmodesmata*). There are no gaps or intercellular spaces in the endodermis; in effect it forms a continuous cylinder or sheath of tissue around the stele. Airspaces in the cortical parenchyma may abut on to it, but it also acts as a barrier to the inward continuity of the gaseous phase, and gases thus can only pass across the endodermis in the dissolved state (aqueous phase).

If the suberisation extends around to the inner tangential wall the endodermis is said to be in the secondary condition. In mono-cotyledonous roots one may frequently (and conveniently) see endodermis in the tertiary state, in which the inner tangential walls of the sheath in continuity with the radial walls are now also cellulose-thickened, suberised and often lignified as well (Fig. 7 · 4c and d). This would result in a total barrier to water movement, were it not for the presence of *passage cells* where the inner tangential wall is unthickened and unsuberised. Because these passage cells are living, they can presumably control the radial passage of water and solutes as well as when they were in the primary condition. A secondary endodermal sheath is in effect like a tin can with holes punched in it. Water can only enter or leave the stele via the ' holes ', namely the passage cells.

In the oldest tertiary stages of an endodermis, all walls of the cells become thickened and lignified, though passage cells may still remain (Fig. 7 · 4d). We need at this point, however, to pay most attention to the primary endodermis, since from the point of view of the working root, this is most likely to be the point of maximum control of water and perhaps solute movement.

7 · 4 *The uptake of water*

From our discussion of the Soil–Plant–Atmosphere complex (section 6 · 3) it can be seen that root cells take up available water, because the loss of water from the aerial parts of the plant induces a condition of water strain, and this is transmitted via the xylem to the roots, and via living root cells to the soil water. If we think in terms of the effective concentration of water (the physical chemist would define this more precisely as the *activity* of water) then it is fairly easy to grasp that water must flow from a region of high activity to a region of lower activity. Loss of water during transpira-tion lowers the activity of water in the mesophyll cells. Water flows from the vascular system towards this region of reduced activity, in doing so the activity of the water in the vascular system itself becomes lowered, and this can be expressed in terms of a developing hydrostatic tension, which is transmitted across the continuous columns of water to the protoxylem vessels of the ‚young roots. At this point the vessels are in contact with living cells, and a gradient of water activity is completed across to the outermost absorptive areas of the very young root. If there is no barrier to the flow of water, and if soil water is abundant, water will move across the gradient, from the more or less saturated leaf airspaces, up the plant and out into the drier atmosphere which surrounds the plant. The sun provides energy for evaporation ; the latent heat of evaporation of water at 20°C is just over 2440 Jg^{-1} of water.

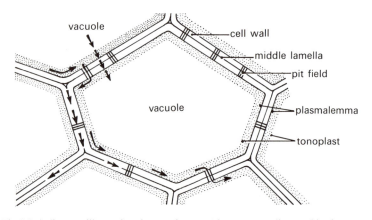

Fig 7·5 A diagram illustrating the apoplast-symplast concept discussed in the text.

Needless to say, the picture is not quite as simple as this. In the first case the choice of route for water flow is likely to be determined by the relative resistance to flow of the pathways that are available. In the young root, water and dissolved substances can move across the cortex entirely within the completely permeable cellulose cell walls. This water-permeated system lies *outside* the living osmotic barriers of the cortical cells, and has been termed the *apoplast*. Water can also flow along an osmotic gradient from cytoplasm to cytoplasm of adjacent cells, across cell membranes that are separated by cell walls. And finally, water can also travel in continuous cytoplasm from cell to cell, via the delicate protoplasmic strands (plasmodesmata) which traverse the walls. This continuous cytoplasmic system has been termed the *symplast* system, and it is clearly involved in the transmission of dissolved substances as well as water inside the osmotic barriers of associated cells (see Fig. 7 · 5).

Earlier work by Lundegardh on the uptake of water by whole root systems in *Helianthus* showed that the amounts transferred fell off with decreasing temperature. This argues that there is a measure of metabolic control at some point along the route, and this view is supported by evidence that respiratory poisons (such as carbon monoxide, cyanide etc.) may also cut down the rate of water uptake by root systems.

It is not only water that passes into the xylem; there are varying amounts of ions (see following section) and non-electrolytes such as sugars. These vary with season, and for example in pear tree sap, the sugar content has been shown to build up to a maximum just

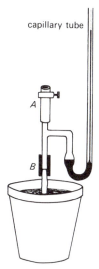

capillary tube

A

B

Fig 7·6 Apparatus by means of which root pressure may be demonstrated. The T-piece and clip at A make it easier to exclude air. The rubber junction at B can be slipped over the freshly-cut stem, using a cork-borer of appropriate diameter. Higher pressures may be demonstrated if mercury is used as illustrated, but water alone may be used in the manometer.

before spring, rapidly dropping as soon as active growth is resumed, and remaining low throughout the summer. It may be presumed that sugars either leak into the xylem from adjacent living parenchyma, or that they are actively secreted there. Whatever the cause, such solutes contribute an osmotic component that could account for the movement of some water into the transpiration stream.

With some pot plants such as *Fuchsia* or *Hydrangea* it is relatively easy to demonstrate *root or exudation pressure*. Figure 7 · 6 shows an experiment in which a recently watered plant of *Hydrangea* is exerting a pressure on a mercury column of nearly one fifth of an atmosphere. This implies that water is being actively secreted into the dead tracheary elements. If there are other shoots on the plant which are transpiring normally, it may not be possible to demonstrate root pressure, and such shoots should be enclosed in a plastic bag in order to reduce transpirational loss.

Here one may reasonably ask at what point does the secretion of water into the vascular system take place. Is it, for example, in the living parenchyma adjacent to the dead but water-filled xylem elements? or, since water does not have to come under full osmotic control until it reaches the continuous endodermis with its controlling suberised Casparian band, should we not rather look to the endodermis as the main point of control? It may be, in fact, that all living cells from the endodermis inwards are concerned with this process. It is certain that exudation pressure, like water uptake, is susceptible to the action of respiratory poisons.

7 · 5 *The uptake of electrolytes*

The absorption of ions by living systems involves passive and physical processes; it may also involve active and metabolic processes. Salt uptake can be studied in a variety of living systems, but the systems chiefly favoured by plant physiologists have included certain large algal cells, discs of storage tissue such as potato and carrot, root systems isolated from their shoots, and the root systems of intact plants.

Studies with storage tissues have made it clear that salt uptake is very dependent upon the physiological state of the tissue. Thus, freshly-cut potato tuber tissue is much less active in accumulating ions than similar tissue which has been ' aged ' by continued washing and aeration for a period of four or five days. After ageing treatment the oxygen consumption of the tissue rises, and clearly salt uptake is dependent upon respiratory metabolism, for it can also be shown that both are speeded up by raising the temperature (see Fig. 7 · 7), and further, that salt uptake can be suppressed by the use of respiratory poisons such as cyanide and carbon monoxide, or by lowering the oxygen tension.

The use of excised root systems confirms these features of salt uptake, and it can be demonstrated, for example, in excised barley roots, that absorption is greatest at a distance of up to about 4 mm behind the apex. As one moves away from the apex to older parts of the root the capacity for ion uptake diminishes. Excised root systems can often be grown satisfactorily in sterile culture, and have the advantage that batches of physiologically similar 'clonal' material can be grown and used for comparative experimentation. Attached root systems are perhaps rather less easy to grow in sterile culture, but it can be shown that the presence of micro-organisms may make a significant difference to the availability of certain ions to the plant (especially phosphate).

The advantages of using single large algal cells are that micro-chemical analyses may be made of the cell sap, and also that by inserting micro-electrodes into these large cells, much information can be gained about the flow of specific ions.

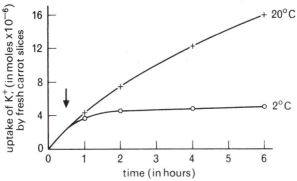

Fig 7·7 The timed course of uptake of potassium (K^+) ions by carrot tissue at two temperatures. Up to the time indicated by an arrow, the uptake is physical and mostly concerned with the saturation of cation absorption sites in the apoplast. After this time, uptake is largely metabolic; it is very slow at 2°C and a good deal faster at 20°C.

Studies of this kind focus attention on the relative roles of plasma-lemma and tonoplast membranes. If a salt-starved cell is immersed in a solution containing ions, equilibrium by diffusion will be reached after a relatively short period. Diffusion is accompanied by adsorption of the ions on to available ion-binding sites within the wall or in the cytoplasm; there may also be some exchange of one ion for another at the membrane boundaries, and in some cases there may be leakage and inwards diffusion across the plasma-lemma. A glance at Fig. 7·8 will show, however, that these physical processes give place to a slower process which is dependent, unlike them, upon temperature; it is a metabolic process which can be shown to be related to the production of energy, mediated through ATP produced during respiration. This process involves a secretion of ions into the vacuole, and, just as work must be done to convey ions inwards across the tonoplast membrane into the vacuole, so also work must be done to get them out.

A number of theories have been developed to account for the uptake of ions by roots. One of them suggests that anion uptake is linked to respiration, and that cations passively accompany the entering anions, thus maintaining electrical balance in the cell. However, there is some evidence that cations such as K^+ may enter in association with highly specific carrier molecules, and some further evidence that there is a positive energy-linked pumping action for anions such as chloride. It by no means follows that the mechanism for entry is the same for each species of anion or cation, and indeed the capacity of a plant for selecting ions may well hinge upon the use made of differing absorptive mechanisms.

It is not appropriate that this complex but fascinating process should be explored here in any greater depth. However, it is useful to have the concept of a movement of ions that can be either along the water-imbibed cell walls, or, via the plasmodesmata from the cytoplasm of one cell to that of its neighbour. The word *symplast* has already been used to describe this idea of cytoplasmic continuity throughout a tissue; it also implies that the vacuoles are isolated by the tonoplast from the neighbouring cytoplasm, and from one another. A vacuole is a kind of store-house or a waste-bin in which materials may be set aside from the other working parts of a cell, and in which the tonoplast acts as a guardian of whatever enters or leaves the vacuole (Fig. 7 · 5).

The cortex of the root system fits well into this picture. Ions may be exchanged between the surfaces of soil particles and root hairs in direct contact; alternatively ions may be displaced from soil particles by H^+ or HCO_3^- ions produced in respiration, and can then diffuse into the root system via the epidermis and root hairs. Once in the root, ions may travel along the cell walls; they may also travel from cell to cell via the cytoplasmic connections or plasmodesmata, or they may be stored in the vacuoles. At the endodermis, diffusion inwards is stopped by the Casparian band, and entry into the stele can only take place along a cytoplasmic pathway. Once inside the stele, ions can be released into the transpiration stream.

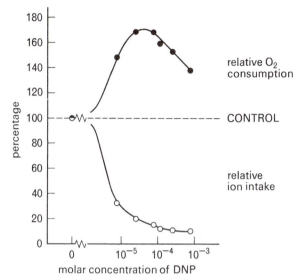

Fig 7·8 The effect of metabolic disturbance on ion uptake. Dinitrophenol (DNP) affects the ability of the mitochondria to make ATP by coupling ADP with inorganic phosphate, using some of the energy made available during oxidative electron transfer (see *Molecules and Cells* p. 127). Because the process of phosphorylation acts as a regulating brake on the respiratory system, one of the effects of DNP is to 'release the brake' and allow the system to race. Under these circumstances the rate of oxygen consumption rises, but there is no ATP available to facilitate ion uptake, so that the rate of ion uptake falls. Oxygen consumption itself starts to fall off as DNP reaches toxic levels.

In this graph, oxygen consumption and ion uptake are both plotted as a percentage, referred to control rates in the absence of DNP, and are related to DNP concentrations.

Studies with excised root systems make it possible to pick up and analyse the exudate from the cut xylem vessels, and it may be observed that a secretion of ions has taken place into the xylem against a concentration gradient; in other words, ions have been concentrated or accumulated from the external solution into the xylem sap. This can be compared with the accumulation of ions into the vacuoles of cortical cells, or into the large algal cells mentioned above.

With these things in mind we can turn again to the root system of the intact plant. Uptake is dependent upon the same factors as operate in excised root systems. Plant roots do not work at their best when they are starved of oxygen, as when they are growing in a waterlogged soil, or when the soil temperature is low, as frequently happens during the winter in northern latitudes. It seems on balance of evidence that the uptake of ions is not as much influenced by variation in transpiration rates as might be expected, though it is obvious that concentration gradients of ions between root and shoot will more rapidly come to equilibrium when the transpiration rate is higher. Whole plant studies of root systems emphasise in addition the effects of translocation. A demand for certain ions exists wherever living material is being synthesised, i.e. especially in meristematic and expanding tissues. Such regions constitute ' sinks ' or regions of demand. By their ion-absorbing activity roots constitute a region of supply, although cell physiologists have not as yet managed to unravel the complex processes by which the shoot makes its demands, either on the ions that are readily available (and can diffuse into the root via the cell walls and the symplast system), or on the ions that may have to be released from vacuolar storage. Furthermore, there is some evidence that ions may move across a corky endodermis, provided that there is living cortical tissue adjacent to it; this is tantamount to saying that the absorbing surface of the root as a whole is not necessarily confined just to the neighourhood of the root tip. Nevertheless, it seems likely that the endodermis and living stelar parenchyma may play a prominent part in the regulation of the ions that enter the vascular stele and move up the plant, but the means by which this control is itself regulated is not by any means clear.

Thus whilst the absorbing surface of the root is not itself the main site of control for the materials which pass up the stele, both root hairs and cortical parenchyma may act as reservoirs for various ions that are required for metabolism and growth, and these may pass via the symplast system across the endodermis and into the stele. At the same time, when under dry conditions the soil water is being depleted and becomes more difficult to obtain, the root hairs and cortical cells may have to perform osmotic work in order to obtain water. If water is not freely available in the apoplast, i.e. the space outside the symplast, an osmotic gradient of traditional pattern may have to be established from cell to cell between endodermis and soil water.

8   The translocation of solutes

The normal processes of supply and demand involve transport: in the working plant the materials that are manufactured in the leaf are redistributed to growing points (e.g. in the active root system), to regions of storage, and during reproduction to the developing flower, all by the phloem transport system.

8 · 1   *Anatomical considerations*

8 · 1 : 1   Transfer cells

Many situations occur in plants in which soluble materials must be transported in quantity across a cell membrane. This happens, for example, in the secretion of nectar from a gland, in the transport of food materials between parenchyma cells and the conducting elements of the vascular tissues, and in the translocation of materials into a young developing embryo within an ovule. In all of these instances large amounts of dissolved materials must pass across a plasmalemma which acts as a barrier to them and slows down the rate of flow. However, in all the situations described above, and in many others where similar requirements for a high rate of solute flow across a membrane occur, specially modified cells are found which have been called ' transfer cells ', and which are thought to assist such a flow. The walls of these cells are extended into finger-like projections (Fig. 8 · 1), sometimes branched and complex, and the plasmalemma follows the surfaces of the projections in the same way that it usually follows a smooth cell wall. The area of the plasmalemma is greater than that of a smooth-walled cell of similar size, and the increased surface/volume ratio facilitates the passage of solutes through these cells.

8 · 1 : 2   Phloem

The distribution of phloem has been dealt with in Chapter 4. It may be recalled that in primary stems phloem tissues run outside the xylem and on the same radius. Occasionally in some families (tomatoes, marrows etc.) further phloem develops also on the margin of the pith next to the protoxylem. In primary roots phloem

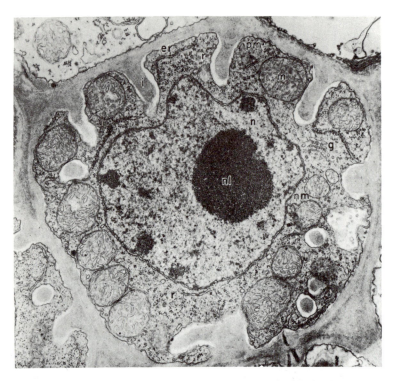

Fig 8·1 An electron micrograph of a transfer cell in a leaf vein of *Galium aparine* (goosegrass) showing ingrowths of the cell wall (i) which increase the area of the plasmalemma (pm) and thus the capacity of the cell for handling the movement of solutes.

Abbreviations: n, nucleus; m, mitochondrion; nu, nucleolus; nm, double nuclear membrane; er, endoplasmic reticulum; r, ribosomes; gb, Golgi body.

groups are found between the ridges of the solid and central xylem core (i.e. on alternate radii to those of the protoxylem groups). During secondary thickening in both stem and root, new phloem is added outside the cambium; the oldest phloem is always to the outside of the secondary phloem and therefore generally under compression. The new phloem is the most active.

c. T.S. of mature phloem in *Cucurbita sp.* The sieve plate (sp) shows a series of dark areas (a) where the cytoplasm has taken up the stain, largely due to the presence of slime protein (cf. Fig 8·4). The lighter areas around them are callose, and these in their turn are separated by cellulose, which is refractive and appears white.

d. L.S. of part of a sieve tube of *Ecballium sp.* (squirting cucumber) showing a sieve plate with its adherent slime plug (slp). This is typical of phloem which is fixed for microscopy without special precautions. The tapering companion cell (cc) ends at the level of the sieve plate (sp); e, sieve element.

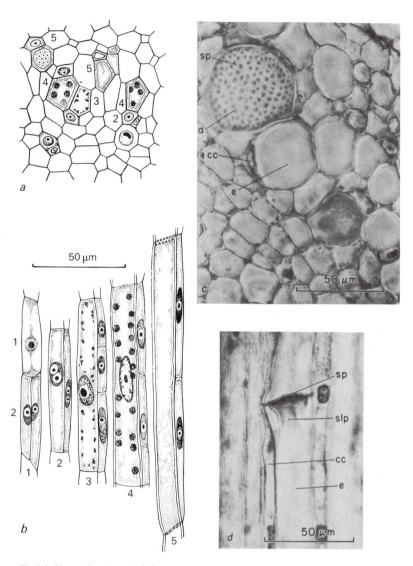

Fig 8·2 Sieve tube characteristics.
a and *b*. Drawings of the stem of *Cucurbita sp.* in T.S. and L.S. showing typical phloem (after Esau). Successive stages in differentiation are numbered 1–5. The cell division which cuts off a companion cell from a sieve tube may be followed by other divisions resulting in more companion cells in a vertical series. The appearance of slime bodies begins in 3 and develops in 4. By stage 5, the slime is dispersed (cf. Fig 8·4) and the walls are becoming thicker and more lustrous. The walls separating the sieve elements and the companion cells bear sieve areas which are penetrated by plasmodesmata. (Continued on p. 112)

Structure and Function in Plants

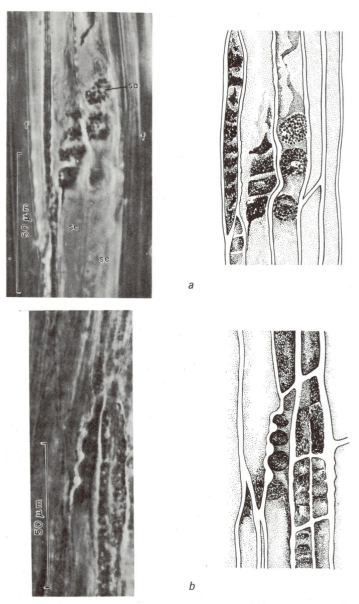

Fig 8·3 Radial (*a*) and tangential (*b*) views of the sieve areas (a) in *Tilia europaea* (lime) which form a compound sieve plate on a steeply sloping end wall. Two sieve elements (e) lie side-by-side in *a*, between phloem fibres (f) on either side.

Of the phloem tissues not concerned with support and protection (i.e. fibres), phloem parenchyma may be concerned with storage and the short-distance movement of food, but sometimes it is part of a radially penetrating ray system, again used for storage and short-distance lateral translocation.

The sieve elements Fig. 8 · 2 are specifically concerned with fast long-distance movement of certain solutes, and their anatomical structure emphasises this special function. The sieve elements are frequently rather thick-walled, and in fresh preparations the fully-imbibed walls appear glistening-white and highly refractive. Certain areas of the walls are penetrated by pores through which pass groups of *plasmodesmata*, protoplasmic strands that connect adjacent protoplasts. These are known as *sieve areas*, and when they occur on the transverse end walls of sieve elements, the latter are called *sieve plates*. The sieve areas may give the wall a ladder-like (scalariform) appearance (Figs. 3 · 6d and 8 · 3). In the flowering plants sieve elements (analogous to vessel segments in the xylem) form continuous *sieve tubes* of considerable length along the plant axis. During differentiation sieve elements nearly always cut off one or more *companion cells* to one side. The companion cells appear to be highly differentiated parenchyma; during differentiation they retain their nucleus and dense cytoplasmic contents, whilst the sieve elements alongside soon lose their nucleus and become vacuolate. It seems likely that the control of events in the sieve element is controlled by the nucleus of the companion cell. Conifers and ferns do not show this differentiation of companion cells, though it is conceivable that adjacent parenchyma may indeed fulfil this function.

8 · 2 *A short note on phloem physiology*

The protoplast of the sieve element seems to be capable of plasmolysis, though its vacuolar boundary is ill-defined and the vacuolar space is in any case longitudinally traversed by cytoplasmic strands. In active phloem, cytoplasmic streaming of small particles can be seen in these strands. There is little confidence in the view that strands (and streaming) are continuous between cell and neighbouring cell; however, Fig. 8 · 4 confirms that the plasmodesmata pores are occupied by cytoplasm and slime protein. Plastids are present, and produce a polysaccharide closely akin to starch. Another

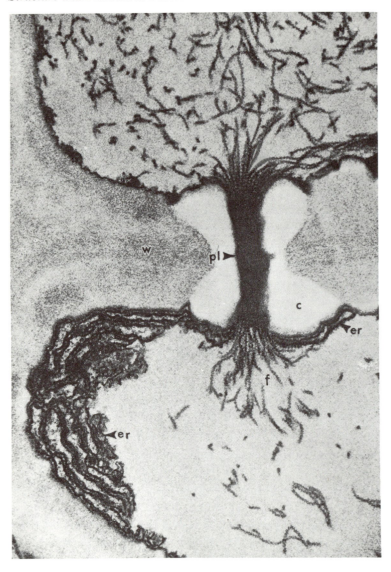

Fig 8·4 An electron micrograph of part of a sieve plate in a phloem cell of *Cucurbita sp.* A plasmodesma (pl) crosses a pore which has been partly occluded with callose (c), which has been added to the margin of the original pore through the sieve plate wall (w). The continuous cytoplasm contains slime protein fibres (f) which have taken up the dense electron stain. Endoplasmic reticulum (er) can be seen in the cytoplasm below the sieve plate.

carbohydrate called callose may be secreted on to the sieve plate in such a way as to close gradually the pores through which groups of plasmodesmata pass. By this means sieve tubes may be sealed off completely during the winter, and only with the advent of the new season is the callose removed by enzymes, and the channels of communication re-opened. Older sieve tubes may never recover from this imposed dormancy, and in the course of time may be crushed and obliterated, their function being taken over by the younger more active phloem.

In view of the discussion in the following paragraphs about the fast movement of solutes in phloem, it is perhaps not surprising to learn that active phloem tissue has a higher rate of metabolism than the parenchymatous tissue that surrounds it. If vascular bundles are stripped out of the petioles of plantain leaves, their respiration rate (expressed as oxygen consumed per gramme fresh mass) may be found to be as high as $800 \, \text{mm}^3 \, \text{g}^{-1} \, \text{h}^{-1}$. Values of $500 \, \text{mm}^3 \, \text{g}^{-1} \, \text{h}^{-1}$ are quite typical. Whole leaves generally show an oxygen consumption of about half this rate, and petioles with their bundles removed (leaving parenchyma only) respire only about one quarter as fast. For contrast and comparison, note that ordinary fresh cut carrot storage root tissue respires at the rate of about $30–40 \, \text{mm}^3 \, \text{g}^{-1} \, \text{h}^{-1}$, whilst active root tissues of maize or sweet corn show a range of respiratory activity from 500 to $1\,200 \, \text{mm}^3 \, \text{g}^{-1} \, \text{h}^{-1}$ according to the distance from the tip.

8 · 3 *Translocation of solutes in the phloem*

Malpighi (in the late 17th century) and Hales (whose *Vegetable Staticks* was published in 1727) both carried out ringing experiments on saplings, in which the bark was totally interrupted by cutting away a ring of tissue down to the cambium. Swelling and sometimes root formation occurred above the ring. If two rings were made, swelling would only occur when there were active leaves left between the two rings. Such experiments may be repeated in the field, or in the greenhouse using cuttings of willow which may very easily be established.

Until the advent of the use of radio-isotopes, the technique of ringing remained one of the main experimental techniques in this field, but with improved analytical methods for the determination of carbohydrates and nitrogenous compounds, a great deal more became known about the movement of organic compounds within the plant. Outstanding in the earlier part of this period was the work of Curtis of Cornell University, and especially the team of workers in Trinidad under Mason and Maskell, who investigated translocation in the cotton plant.

Some of the facts that emerged during this period can perhaps best be summarised as follows:

1. When leaves are active on a ringed stem, organic solutes such as sugars accumulate in the tissues above the ring and are depleted below it.

2. Sucrose is the main form in which carbohydrate is transported, although some variations from this are noted on p. 120.

3. The passage of sucrose is restricted to the phloem, although there may be radial gradients away from the phloem of other mainly hexose sugars, e.g. glucose.

4. Sucrose travels from a 'source', e.g. active green leaves in the light, to a 'sink', e.g. a growing root tip, or an expanding potato tuber; sucrose travels from regions of high concentration down the gradient to regions of low concentration.

5. The direction of movement depends on the relative positions of sources and sinks. On a leafy shoot, if illumination is restricted to the lower leaves, they will produce sugars which move upwards to supply darkened leaves above. Sugars move upwards from the flag leaf at the base of a developing wheat spike into the florets and subsequently the grain of the developing spike.

6. The rate of sugar movement may be increased by diminishing the area of phloem tissues across which it has to pass. This is rather like putting one's finger over the end of a hose gushing water; the water comes out at a faster rate under the pressure which builds up following constriction.

7. Sucrose moves very fast in phloem tissues; its rate of movement is measured in $g\,h^{-1}$ and is not to be confused with its velocity measured in $cm\,h^{-1}$. The latter concept seems to imply that there is a bulk flow of sugar solution through the plant, and that the speed of movement of its advancing front can be measured in terms of the volume which crosses a known cross-sectional area of tissue in a given time, viz. cm^3 per cm^2 per hour or $cm\,h^{-1}$. If the movement of sugar is not by means of bulk solution flow, then obviously the first concept, though more difficult to measure, is more valid. The observed rate in cotton plants is of the order of 4×10^4 times as fast as can be accounted for by the free aqueous

diffusion of sucrose. Characteristic velocities of sucrose movement in a variety of plants range from 40 to 100 cm h^{-1}.

8. The movement of solutes is dependent upon metabolic energy, and is depressed in the absence of oxygen, or in the presence of respiratory poisons. If the petiole of a leaf is chilled, thus depressing its respiration rate, the export of sucrose from the leaf lamina is arrested, and more starch may be detected in the lamina by comparison with a control. Similarly (see Fig. 8·5), in tomato

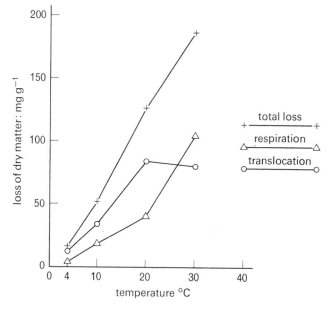

Fig 8·5 A graph showing the effect of temperature on the loss of dry matter from the leaves of a plant of *Solanum lycopersicum* (tomato), over a thirteen-hour period in the dark.

plants that were held in darkness for a 13-hour period, leaf samples showed a total loss of dry matter, which increased with increasing temperature over a range from 4°C up to 30°C. Above 20°C, translocation fell as reserves were used in respiration. At low temperatures, the overall respiration rate (and thus metabolism) was reduced, and movement of dry matter was correspondingly low. Translocation requires some of the energy that is made available through respiration, but both processes draw upon the same pool of metabolites.

8 · 4 *The use of radioactive tracers*

The introduction of labelling techniques using radio-isotopes has helped considerably to clarify and confirm the previous records of high velocity transport, and has made it possible to demonstrate simultaneous two-way movement of different solutes in the phloem. However, we have not yet achieved sufficient technical finesse to be able to demonstrate two-way flow in a single sieve tube. Nevertheless there is now fairly convincing evidence that different substances may travel at rather different speeds in the plant, and it may be presumed that they are travelling independently of one another.

Another very useful technique due to Zimmermann makes use of the fact that when aphids feed, their stylets penetrate most accurately into the host phloem, and sugar is forced up the stylet by the hydrostatic pressure which can be shown to exist in the sieve tubes. The willow-feeding aphid (*Tuberolachnus salignus*) has been used most frequently, and small colonies are allowed to feed at predetermined loci down the stem; their stylets penetrate into the phloem and, whilst feeding, the aphids are narcotised with CO_2 and are severed from their stylets with a sharp blade, leaving the inserted stylets in position. Radioactive $^{14}CO_2$ is assimilated by a leaf or group of leaves near the apex of the host shoot, and labelled sucrose and other products of photosynthesis are passed down the sieve tubes. Sap from the sieve tubes is exuded as droplets from the cut end of the stylets. The droplets can be removed and analysed by chromatographic and autoradiographic methods.

Thus it is confirmed that only very small quantities of hexose sugars are normally found in the sieve tubes. Sucrose is usually the main sugar of transport, though in some plants other oligosaccharides, such as the tetrasaccharide stachyose found in American ash, have been reported. The sieve tubes in addition show the presence of amino acids also in transit. Measurements of radioactivity serve to confirm that the velocity with which the sucrose 'front' advances down the stem is of the order of $100 \, \mathrm{cm \, h^{-1}}$ in willow.

The technique of autoradiography is useful in addition to its role in identifying substances during chromatography. For example labelled solutes that are introduced into the plant and translocated elsewhere, can later be detected by the effects of radiation from the treated material upon X-ray film. For this purpose the plant is quickly killed and dried, and laid out flat in close contact with a lightproof envelope containing the sensitive film; this is exposed to radiation for a number of days, which has to be varied with the ' hardness ' and intensity of the radiation.

Light is thrown in this way upon a number of situations. For example, it is possible to show that once ionic calcium has passed up a stem in the transpiration stream, it appears to be held in relatively immobile form in the leaves (Fig. 8 · 6a and b). By contrast phosphorus applied as $H_2{}^{32}PO_4{}^-$ (8 · 6b), is mobile in the plant, and is monopolised by the youngest leaves and shoot apex, being moved up from older leaves as the plant grows on (mostly in combined organic form). The two sets of drawings based on photographs (Fig. 8 · 6a and b) also show that it is possible, by exposing the plant for different times to solutions containing labelled material, to obtain data concerning the approximate rates of movement of solutes about the plant. It need hardly be said that the photosynthetic incorporation of $^{14}CO_2$ into a leaf and its movement down a stem to a point where it accumulates above a ring provides an excellent method for checking more meaningfully on the observations made so many years ago by Malpighi and Hales.

8 · 5 *Mechanism of translocation*

A good deal of dispute has arisen from the attempts of physiologists to explain the rapid rates of movement of materials in the phloem. Most elementary textbooks pay tribute to the mass-flow hypothesis, but the fact is that no one theory has yet succeeded in convincing the critics of its total aptness to account for the facts. The facts that have to be accounted for, in summary, are:

1. Hexose sugars made in the leaf mesophyll have to be ' loaded ' in the form of sucrose into the sieve tubes.
2. Solutes in the sieve tubes move at high velocity, and apparently independently of one another.
3. There is a positive hydrostatic pressure in the sieve tubes which under favourable conditions results in considerable fluid exudation when the phloem is severed.
4. Phloem has a high rate of metabolism which is susceptible to metabolic poisons, oxygen lack etc.
5. Anatomically, phloem presents in its sieve areas altogether too small a cross-sectional area to allow for fluid flow. Calculations

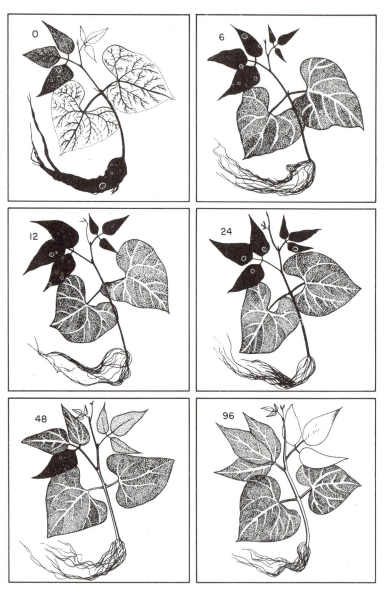

Fig 8·6 Radio-isotopes may be used to compare the relative mobility of selected ions in plants. In one experiment, bean plants (*Phaseolus vulgaris*) were allowed to absorb labelled calcium (a) or phosphate (b) for one hour. They were sampled at 0, 6, 12, 24, 48 and 96 hours after this treatment, having first been removed to a non-radioactive solution.

(a) Calcium moves out of the root system fairly quickly and accumulates in the terminal leaves; there is no appreciable subsequent translocation into the newest and youngest leaves.

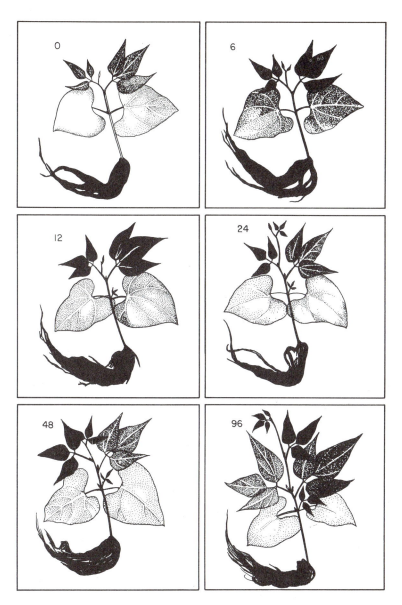

Fig 8·6 (continued)

(b) Phosphate pervades the root system, and passes quickly to the active terminal leaves. It remains sufficiently mobile to be moved to younger leaves which develop later.

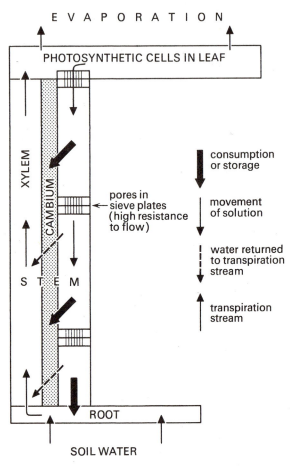

Fig 8·7 A diagram to illustrate some aspects of mass-flow discussed in the text.

show that absurdly high pressures would have to operate to account for the flow of fluid across cytoplasm-filled pores.

The mass-flow hypothesis postulates high osmotic potential in the leaf phloem ('source') and low osmotic potential in the regions of consumption or storage ('sinks') where, also, water is removed to the xylem (Fig. 8·7). According to this theory, water is taken into the sieve tubes from the transpiration stream, and hydrostatic pressure mounts, leading to bulk export down the sieve tubes. As thus visualised, it is a physical system that is independent of metabolic energy, except maybe in the 'loading' process. Yet, as we have seen, translocation out of a leaf may be very severely hindered by local chilling of the petiole, some centimetres away. This surely suggests that the transport system itself is driven by metabolic energy, or that energy is required to keep the channels open for bulk transfer.

There seems little doubt that the process is somehow geared to the provision of metabolic energy, and in view of the high respiratory rates of phloem tissue that have been reported, this can presumably be made adequately available. All but the more elderly sieve tubes show some active cytoplasmic streaming; this may be because more energy is available, or it may be connected with solute transfer. Observations on streaming in the so-called transcellular strands are still the subject of controversy, but if substantiated might well provide one means of explaining how solutes may flow in solution across the cytoplasm-packed pores in the sieve plates. However, solution flow does not explain how it is that quite similar molecules travel at different speeds; for example in soya bean, glycine and asparagine travel more slowly than alanine and aspartic acid. Here we may be dealing with different channels of transport, or with the activity of a fairly specific carrier mechanism.

Enough has been said to indicate that we are in no position to make a definite decision in favour of any one particular hypothesis; as with most similar scientific situations, we have to wait until our technical ability is good enough for the purposes of investigating the questions that most need to be asked.

9   Leaf fall

9 · 1   *Anatomy and physiology of leaf abscission*

Plants show a variety of adaptations to the particular ecological niche that they occupy. In dry hot conditions leaves show xerophytic characters by means of which water loss is endured or resisted. In the cooler, moister conditions of our temperate climates, winter brings more or less frost and snow, and less light because the sun is further away and lower down on the horizon. Winter is a period when less photosynthesis and other metabolic activity is possible; if the ground is cold, water uptake will be reduced; if it is frozen, water will not move at all. With the approach of these conditions annuals die off and perennials prepare to overwinter, showing a variety of devices for protecting the growing points within their buds. Of the onset of bud dormancy, and its heralding in the summer months (when conditions are most favourable for growth) nothing will be said here, other than that most of our deciduous or leaf shedding trees have already laid down their buds by July or August. It is well known that they shed their leaves in October or November, and by so doing they reduce the exposed transpiring surface of the tree; water loss is restricted to lenticels, and is very limited in amount.

The changes by which this is brought about are fairly well known, though there is variation from species to species (Fig. 9 · 1). In most woody trees the leaves are borne on twigs of one season's growth with a barely established periderm about to extend forward towards the apex from the periderm of last year's growth. At the base of the petiole a layer of cells becomes differentiated as the *abscission zone*, extending in from the petiole and finally forming a plate of tissue in which meristematic divisions may occur. On the stem side of this plate the cells tend to become suberised and corky. These cells will form the *protective layer*. As leaves age they show signs of degradation (for example solutes leak more easily out of the leaf cells) and they are less metabolically efficient. The likelihood for most plants is that leaf auxin production meanwhile declines and, as it fails, so some or all of the following changes may occur in the abscission layer: (*a*) the middle lamella which separates adjacent cells may degenerate;

(*b*) the intercellular spaces of the tissue may become injected with fluids (due to further permeability changes); (*c*) there may in some cases be mucilaginous degeneration of the walls themselves, and presumably enzyme hydrolysis is taking place. In the experimental plant *Coleus*, the whole process can be speeded by cutting off the

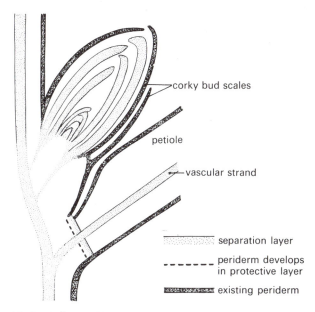

corky bud scales

petiole

vascular strand

━━━━━ separation layer

- - - - - - periderm develops in protective layer

▨▨▨▨▨ existing periderm

Fig 9·1 A diagram illustrating abscission in a woody plant.

lamina, leaving only a petiole stump which will fall off in about three days. Its fall may be deferred by as long as five days if a paste of aqueous lanolin containing an auxin such as IAA (see Section 13 · 1) is smeared on the cut end of the attached stump. This indicates that a diminishing auxin content allows the development of the abscission layer whilst the maintenance of a higher auxin level inhibits abscission. *Coleus* seems to be an example of a relatively simple abscission mechanism, for there are certainly other more difficult events to explain in other species. As will be seen later (Chapter 15) there are a number of events during reproduction in which abscission plays an important part.

In the leaf itself the onset of senescence brings about a fall-off in the chlorophyll content of the leaf and this is accompanied by an export of protein nitrogen and of ribonucleic acid. Most of the mobile ions such as potassium, phosphorus, magnesium and so on, pass out, either to be mobilised in the direction of younger and more active tissue or to be stored in the ray tissue of the wood, or wherever else storage takes place. Non-mobile ions such as calcium remain behind, sometimes in combination with metabolic by-products such as oxalic acid. The green chlorophyll often has been masking accessory pigments such as the carotenoids and the sap soluble anthrocyanin pigments, and its breakdown and removal leads to a display of autumnal leaf colours which is characteristic of the fall. There is absolutely no evidence for the view sometimes stated that during this time *waste* products are passed into the leaf.

Meanwhile the vascular tissues have gradually become blocked off. Living parenchyma cells known as tyloses invade and block the xylem cavities from neighbouring living parenchyma. Callose is deposited on the sieve plates and sieve areas of the sieve elements, and transport of water and solutes finally comes to a standstill. As the leaf dries the abscission layer becomes completely ruptured and the leaf is left hanging only by its vascular bundle, now wholly at the mercy of the winds and the frosts. In the stem behind the corky protective layer the periderm may extend and offer further protection, in this way reducing water loss and invasion by organisms.

9 · 2 *Nutrient cycles associated with leaf fall*

Something must be said about the fate of leaves which fall to the ground and constitute the layer of litter on top of the soil. The processes whereby this litter is converted to humus may be stated simply as follows. Firstly, by the leaching action of rain and the quick invasion of microbial organisms, simple molecules such as soluble sugars and amino-acids are quickly removed from the cells. The next wave of saprophytes include starch- and protein-splitters, and they leave behind skeletons of cellulose, ligno-cellulose and cuticularised material. Cellulose-splitters follow on in the succession and the final slower and more difficult conversion of ligno-cellulose results in residues of humus material (e.g. humic and fulvic acids) together with fatty residues from cuticular materials. In all of these activities fungal and bacterial organisms predominate, assisted by soil animals, e.g. springtails, free-living nematodes and so on.

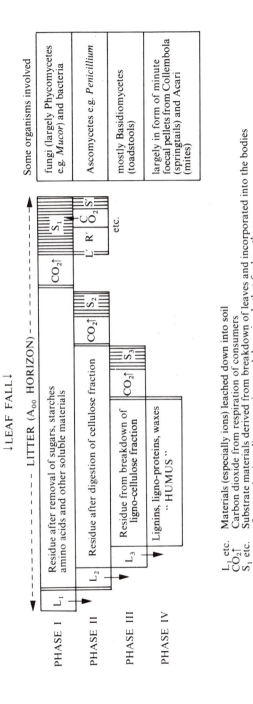

Fig 9.2 A diagram to illustrate stages in the decomposition of leaf litter.

L₁ etc. Materials (especially ions) leached down into soil
CO₂↑ Carbon dioxide from respiration of consumers
S₁ etc. Substrate materials derived from breakdown of leaves and incorporated into the bodies
of successive invading organisms and the animals that feed on them

The process can be depicted diagrammatically (Fig. 9·2) and it will be seen that the dry mass of the original leaf material is diminished, partly by leaching, partly by its incorporation into the living cells of the invaders, and partly on account of carbon dioxide that is lost during the respiratory activity of the micro-organisms.

The rate at which these activities go on may be regulated especially by soil temperature and soil aeration. It is noticeable that a waterlogged soil tends to accumulate organic matter because it can only be converted very slowly to humus residues. A peat soil, as is well known, has high levels of organic material for this very reason.

More important from our present point of view is that during the conversion of litter, there is a gradual release of ions into the soil. Some of these ions are incorporated into the cells and bodies of the living organisms present, but some are available, either in the soil solution or adsorbed to the surfaces of soil particles, for uptake by the roots of higher plants.

Thus we find a circulation of some species of ion as between plant and soil, and this is complementary to the transport of yet other

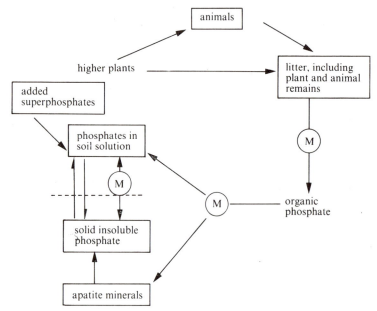

Fig 9·3 A diagram illustrating aspects of the circulation of phosphorus between plants, animals and the soil; Ⓜ indicates microorganisms.

species of ion back into the stem prior to leaf abscission. Elements that are trapped in leaf tissues (such as the not very mobile Ca^{2+} ion), tend to be circulated in a cyclic fashion between plant and soil. Elements such as nitrogen or phosphorus may either be conserved within the plant prior to leaf fall, or they too may take place in a cycle of transport between the soil and the plant. Such a cycle is shown for phosphorus in the adjacent diagram (Fig. 9·3) and the nitrogen cycle offers a similar example of a circulation between plant and soil.

It is perhaps not sufficiently well realised that this intimate balance between soil, soil organisms and higher plants locks up a great wealth of nutrients which supports all the members of the biome. It is a recognised fact, for example, that the clearing of luxuriant tropical forest leads to considerable leaching losses of the available nutrients that are in circulation in the soil–plant complex. The fertility of the soil is seriously depleted, and it takes a very long time for soil, soil organisms and vegetation between them to establish once again the same rich intimacy of nutrient exchange.

PART THREE

Growth, differentiation and development

10   The general physiology of growth

10 · 1 *Measurement of growth*

In order to assess growth in plants we have to adopt a particular view about each particular aspect of growth. Whereas a forester may define the growth of a tree in terms of its increase in height and girth, a farmer must inevitably think, for example, in terms of yield per acre, and even this would mean different things for different crops. It would be measured for wheat in bushels per acre or kilograms per hectare, but the tonnage of sugar beet would mean little unless one also had some idea of the percentage sugar content of the storage root. Then again, brewers and bacteriologists might assay the growth increase of their cultures by counting the number of cells in a given small volume of fluid; since well-shaken cultures of yeasts and bacteria are usually fairly uniform, this is as good a method as any for following the growth in size of a population of this kind.

The assessment of growth by counting at intervals the numbers of individuals in a given population is satisfactory in a yeast culture because the process of increase involves budding and fission, followed by increase of cell substance to the point at which division occurs once again. If we centrifuge the colony down into a close-packed mass and weigh the pellet, we shall clearly have a measure of the *fresh weight* of the total number of cells present, though we shall also weigh some extra water that occupies the interstices between individual cells. If we dry the pellet in an oven under standard conditions until its *dry weight* remains constant, we shall have eliminated the variation due to the water present. We could go further, for example, and chemically estimate the total nitrogen present in the pellet, this being a useful way of characterising the amount of living material present since protein may account for as much as 40% of the dry weight of protoplasm. The objection might still be raised in yeast that this takes no account of the varying amounts of storage carbohydrate (glycogen) present which are not necessarily related to the nitrogen content, but do relate to the particular conditions under which the yeast was cultured. There is clearly no one satisfactory overall criterion for measuring growth and we therefore deliberately select a criterion that serves our immediate purpose best.

Note also that whilst the yeast cells in a culture may be weighed fresh and then re-dispersed in culture solution to go on growing, their dry weight and their nitrogen content cannot be ascertained without killing them. For purposes such as this it is necessary to take *samples* of the whole culture whilst it is evenly suspended in culture solution. The sample of cells is then centrifuged down to separate the cells from the bulk of the culture solution; these are washed by suspending in ion-free water; they may then be spun down and rewashed until free of contaminants, and then finally spun down to a pellet for drying or for chemical analysis.

In the case of higher plants, the position is further complicated by the fact that increase in cell number is largely (but not wholly) restricted to specific regions known as meristems; these in other words are regions where mitotic frequency is high. In the zones adjacent to meristems, emphasis is on growth by increase in cell size, together with the changes in the cells themselves which we call differentiation. Mitotic frequency falls off as the tissues are left further and further behind the growing point. Now this means that when we estimate total growth in higher plant tissues, we are lumping together zones of intense mitotic activity with zones of maturing (and matured) tissues. Furthermore, some cells of the matured tissues (e.g. xylem) will have no living contents. Other tissues will be concerned with active photosynthesis (chlorenchyma); storage tissues for example may contain a lot of starch. The whole plant body presents a complex of dispersed activity which makes the measurement of some aspects of growth extremely difficult.

Generally we may say that the measurement of fresh weight is most useful when we wish to follow continued growth, as when we are growing plants in an aerated water culture. Each plant can be blotted free of superfluous moisture, rapidly weighed and replaced at once in the culture vessels. If we require more information than this, then we have to grow a population of such plants, sample them at appropriate intervals (preferably with statistical control), determine their fresh weights, and then oven-dry them (alternatively freeze-dry them), and then proceed with whatever analysis has been decided upon. The dry weight includes the whole complex of assorted tissues, some alive and some of them dead; once again, determination of total nitrogen gives us a better picture of the amount of living cell substance that is present, but includes nitrogenous storage reserves. Indeed for some purposes, perhaps the best measure of all involves an assay of the nucleo-protein present.

Growth can be seen to involve irreversible increase in cell size and cell number. This also involves irreversible increase in mass and it is important that we take note of how this increase in mass is achieved, whether by increase in cytoplasm, by the storage of reserves, or by the laying down of materials (such as cellulose, ligno-cellulose, etc.) that are mostly unavailable for further syntheses.

It follows that we have to adopt as a criterion of growth that which best suits the experimental conditions, and does not give too false a picture of what is happening, at least for the purposes required. Factors which contribute to variation have also to be reckoned with and controlled. For example, it is useless to adopt leaf area as a measure of growth unless it is fairly certain that the samples of the leaf 'population' being measured are comparable in water content. Thus, in a classical experiment on photosynthesis, sample discs are punched out from large leaves at the beginning of the day, and again later, after a period of illumination. The changes which are shown to occur are not only the result of photosynthesis, but may also reflect the contraction of the leaf under water stress, and may therefore modify the measurements that have been made on weight increase following photosynthesis.

10 · 2 *Control of external factors influencing growth*

Nowadays it is customary to grow individual plants or populations of plants under growth conditions that are as closely controlled as they can be. To achieve this the plant physiologist requires to be able to control temperature, the quantity, the duration and the quality of incident light, the humidity and gas content of the atmosphere, and finally the supply of water and nutrients. A rather expensive way of doing this is to build a 'phytotron' or plant growth chamber, in which all these factors are amenable to control. Temperature is usually controlled by circulating air across heating or cooling systems and humidity may be controlled in a similar manner. Time-switches control the length of exposure to light or darkness. Light intensity is generally measured in terms of energy incident on unit area of the plant surface. It is customary to use fluorescent tubes for most purposes, and some modification of the predominant wavebands emitted by these is possible. Mercury vapour lamps are used when greater intensities are required, but bring with them attendant problems of extra heat production and limited wavelengths of light. The use of cine filters makes it possible to experiment with different, generally rather broad, wavebands of light. For narrower bands monochromators are used, or a continuous spectrum is produced using special prisms and an intense beam derived from a suitable source (see page 199 in discussion of phytochrome). This exemplifies well the lengths to which experimentalists must go in devising suitable controls for their experiments; it is obvious that in order to examine the effects of single factors it is always desirable to control as many of the accompanying variables as possible.

Mineral nutrition is often controlled by growing plants in aerated water cultures. The solutions used have the advantage of being easily changed or replenished, but even when their aeration is very efficient, it can hardly be claimed that submerged root systems are properly comparable to root systems growing in soil. Nevertheless soil cultures are themselves amongst the most difficult to control; the nearest artificial equivalent is washed and sterilised sand or vermiculite. Using such materials it is possible to control pore space and soil particle size, but the whole system makes a pale substitute for a soil with its teeming populations of micro-organisms. The essential point to note however is that if a plant physiologist requires to work on the effects of a particular factor on a plant, he may often be forced to simplify the whole system in order that he may achieve better control of the remaining complex of factors.

11 Differentiation

11·1 *Use of tissue cultures for studying growth and differentiation at the cellular level*

Plant physiologists have now learnt to grow tissues under cultural conditions that make it possible to obtain and maintain suspensions containing single and clumped cells. Tissues from which cultures of this kind have been derived are many and various; they include the secondary phloem of carrot storage root tissue, and a number of stem tissues including the pith of tobacco stem and the cambium and phloem of various woody plants. It is possible with such cultures, by following increase in fresh weight, or by determining change in cell number, to determine the nutritional conditions best suited to their division and growth. A ' basal ' medium is generally supplied; this contains a carbohydrate energy source, such as sucrose, a properly balanced selection of inorganic ions, plus appropriate supplies of soluble reduced nitrogen. (Casein hydrolysate is commonly used, for it provides a range of the requisite amino-acids.)

Most tissue cultures can be helped towards better growth by treating them as if they were *heterotrophic*, i.e. nutritionally dependent on external sources. This is true even of tissues that contain green plastids and could grow *autotrophically*. Some cultures may have only a requirement for ' basal ' medium nutrients; they could be self-sufficient so far as the manufacture of more complex growth materials. In other cultures it is necessary to provide some of these growth materials in small quantities if good growth is to be stimulated and sustained. This is generally achieved by adding coconut milk. It might be expected that any medium in which a plant embryo can develop would contain growth-promoting substances; and it is found that the use of coconut milk in conjunction with basal medium can stimulate even the highly vacuolated cells of tobacco stem pith into active cell division.

In Cornell University, Steward and co-workers have made a thorough study of the constituents of coconut milk, and other similar fluids from immature chestnuts and walnuts. It is found

139

that there is always some form of auxin present (e.g. indole acetic acid, maybe in combined form), though in trials some synthetic auxins are sometimes found to be even more effective; then there is a second group of substances, called the *neutral* fraction and involving sugar alcohols such as inositol. A third *active* fraction contains a range of substances. Amongst these are the kinins (related to adenine) and here again it is found that synthetic substituents are often much more effective in promoting growth. What is most interesting is that the degree of combined action displayed by these compounds is often greater than would be expected from a mere summation of their individual activities.

The term *synergy* is used to describe the situation where two substances acting together produce a more marked effect than can be attributed to their individual effects when acting alone. Synergy is demonstrated in the following figures (Table 11 · 1), based on work from the Cornell laboratories. Here, the effects of various additives in various combinations is tested on the total growth of cell masses taken from carrot phloem, and it will be seen that more growth than expected is found in three out of four of the combinations used. It should be noted that the neutral fraction (inositol etc.) enhances the effect of the active fraction, but diminishes the effect of the IAA. It is known that cytokinins always work best in the presence of IAA, and this is well demonstrated by the figures, which show a synergistic increase of more than 100% over and above the effects of these substances considered individually.

Table 11 · 1

Synergistic effects between various combinations of IAA, neutral (NF) and active (AF) fractions from coconut milk on growth of carrot phloem explants.

Growth Medium	Mean Final weight expressed as % of (1) (a)	Growth Increment % (b)	Expected growth increment assuming additive effects (c)	Increase or decrease % (b)−(c)
(1) Basal medium only	100	—		
(2) BM + IAA	210	110		
(3) BM + AF	132	32		
(4) BM + NF	189	89		
(5) BM + AF + NF	380	280	32 + 89 = 121	+159
(6) BM + IAA + NF	271	171	110 + 89 = 199	−28
(7) BM + IAA + AF	343	243	110 + 32 = 142	+101
(8) BM + IAA + NF + AF	425	325	110 + 32 + 89 = 231	+94

The small fragments (or explants) of tissue used for these tests may be stimulated into callus growth in which the cells that are formed do not separate but proliferate in a mass. Such cell masses do not normally revert to the production of organs such as roots. However, it is possible from these masses to obtain ' shake ' cultures of separated cells, and it is interesting to see how these cells, freed of the constraints and controls imposed upon them by adjacent tissue masses, develop and behave far more like developing embryos, with which they share many features. Most important is the fact that, as they grow and themselves produce aggregates of cells, evidence appears of differentiation of growth centres from which root and shoot apices may be derived from a single cell. The Cornell school have managed to grow such cell aggregates right through to adult flowering carrot plants. This is a triumphant demonstration of the so-called *totipotency* of each individual cell, and it establishes that all cells have within them the total genetic potential, i.e. the genotype, for the development of all aspects of the mature plant. It leaves countless questions unanswered, e.g. how can an unspecialised mass of cells give rise to cells of so many different types? Is this differentiation of cell types dependent upon external as well as internal factors? How can one account for the time sequence in development over the life cycle of the plant? How do processes of cell differentiation become integrated and expressed in organ formation? And so on. . . .

Individual cells are presumed to differentiate by reaction within a micro-environment, which evokes a particular response from their genome. Such a micro-environment may be something as simple as an oxygen diffusion gradient; or it may be a great deal more complicated in terms of a changing pattern of hormonal influences reaching each cell from neighbouring centres of production. Some light is thrown on this by work carried out by Skoog and his colleagues at Wisconsin. They repeated some earlier observations by the German physiologist Haberlandt, and were able to confirm that cell division could be promoted in a block of pith tissue by the expedient of growing a fragment of vascular tissue on top of it. He had earlier shown that adenine (the purine fragment of ATP that is also part of the co-enzymes NAD and NADP) could bring about bud-growth in callus cultures from tobacco pith; its derivative kinetin is even more effective in this respect. There appears to be a balance between adenine and auxin, for in an

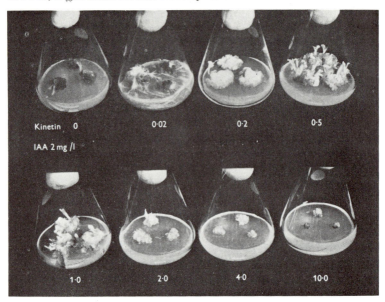

Fig 11·1 The results of an experiment by Skoog and Miller in which explants of *Nicotiana tabacum* (tobacco) callus were grown in sterile cultures on agar, with a constant level of IAA (2 mg l^{-1}) and a variable level of kinetin. Root growth was prominent at a low kinetin level (0·02 mg l^{-1}) but buds grew when the kinetin concentration was increased to 0·5 mg l^{-1}, although its effectiveness fell off at higher concentrations.

experiment in which the basic culture medium was also provided with these substances in varying ratio, a control block of tobacco callus remained unchanged, but if the medium contained 40 µg l^{-1} of adenine, buds differentiated but no roots. The addition of adenine at the same concentration with 1 and 5 µg l^{-1} respectively of IAA produced increasing development of a rooting system. (See also Fig. 11 · 1 which illustrates a similar experiment in which the IAA concentration was held constant whilst the level of kinetin was progressively increased.) So here we have a system in which not only may the development of cells be influenced, but also the initiation of organs. Indeed if we think from the cell towards the organ and thence to the whole plant, the complexity of the whole situation becomes even more acutely apparent. *Organogenesis* is largely controlled through localised activity within meristematic masses. As cell division proceeds, tissues heap up to form papillae at the margins of the dome of the stem meristem; each papilla signifies the emergence of a leaf. Differentiation of the complex of tissues within the leaf (not forgetting the control of its flattened and dorsiventral form) originates partly from the meristems of the leaf itself, but partly also from hormonal and other influences from the shoot below. The shape of the whole plant and the way in which it branches is often controlled by the dominance of its apical bud, which may more or less suppress the development of buds below it, and hence of lateral branches. The growth of a plant involves a high degree of integration between the activities of its separate organs, and these same separate organs have themselves developed and differentiated according to a carefully regulated pattern of events in the apical meristems. We shall briefly discuss these meristems in 12 · 1.

11 · 2 *Aspects of differentiation in cells and tissues*

A meristematic cell grows by the synthesis and incorporation of new materials into its cytoplasm, by the intake of water, and by building new wall materials into the extending cell wall. In doing so the cell increases in volume, and may change in shape and proportions. Above all it may change the overall direction of its metabolism. By this we mean that, perhaps as a result of change in its local micro-environment, new sets of enzymes are being programmed on its ribosomes, and new processes are being carried out. This may become evident in the appearance or increase in number of specific organelles. A cell that is destined to be green and photosynthetic develops chloroplasts; if it were merely going to store starch and not manufacture it on the spot, the proplastids would only develop as amyloplasts. Other switches of metabolic direction are often only detected by their chemical end products, which equally imply changes in the balance of enzymes. Complexes such as lignin and suberin may appear in the walls of certain cells and not others; their development may herald (but not necessarily so) the onset of the death of such cells (e.g. in the differentiation of xylem vessels and cork cells).

Hand-in-hand with the metabolic differentiation of plant cells goes the process of vacuolation and water uptake. Cells of the meristem possess varying numbers of small vacuoles in their cytoplasm; in some cases they appear to develop by a ballooning up of parts of the endoplasmic reticular canal system, and it is by a joining-up and an extension of this early vacuole system that more mature cells come to show the extensive vacuolation that is characteristic of plant cells, and is easily visible under the light microscope. Characteristically as each cell extends in length and in girth, water is taken into the vacuolar system, and the cytoplasm becomes almost entirely peripheral, lying just within the cell wall. Patterns of tissue differentiation are often associated with the differential development of vacuolating tissues, and as will be seen later, the presence of early stages in vasculation is often made more apparent by the greater vacuolation of the ground tissues which surround the pro-vascular strands.

Another aspect of differentiation involves the relative dimensions of cells. For example, anatomists sometimes distinguish between the tissues known as *prosenchyma* and *parenchyma*. In the former tissues the cell initials rapidly develop to long slim cells, much smaller in cross section than their surrounding neighbours, and much more extended in an axial direction. Such cells may stay unlignified, and by laying down extra wall material act as the long supporting cells known as collenchyma; they may later become lignified as well. By contrast, parenchymatous cells remain approximately isodiametric, though most cells show some elongation in the axial direction (i.e. along the direction of the long axis of the stem or root).

Yet another aspect of differentiation concerns the locus of differentiation of the various types of cells. The problems of differentiation are amongst the most intriguing that face us to-day, for not only do we greatly desire to know how it is that a sieve tube comes to differ from say, a pith cell, or a fibre from a palisade cell, but we urgently wish to know what it is that brings about differences between neighbouring cells, for example between phloem and phloem fibres, between vessels which die and neighbouring ray parenchyma cells which live on, or between endodermis and the cortical parenchyma adjacent to it.

Furthermore the relative position, development and rates of growth of various tissues that are laid down in meristematic areas, give form and substance to the organs which develop from them. It is the relative growth rates of adjacent lots of tissue that is responsible for the serrated edges of adult elm leaves, or for the sinuate form of leaves of the durmast oak. Similarly it is the balance between the axial elongation and the transverse expansion of parenchyma cells which determines the shape and proportions of a vegetable marrow or of the more queerly shaped ornamental gourds. We are indeed a good deal in the dark as to how these and many other complex *morphogenetic* processes are controlled.

12　Apical meristems

12 · 1　*Apical meristems and patterns of tissue differentiation*

Growth results from increase in cell number and a subsequent increase in cell size. It has already been emphasised that mitoses yielding more cells are mainly but not wholly confined to meristematic zones. In Chapter 4 we saw that the increase in girth of a root or stem during secondary growth is due to the activity of *lateral* meristems. In what follows we shall be primarily concerned with the activity and contributions of apical meristems.

12 · 1 : 1　Root apex

This has been considered briefly before (p. 98) but will be used as an example of an apical meristem, because it has the advantage of being uncomplicated by the emission of lateral organs, viz. leaves and buds, as in the stem apex. The intense mitotic activity that takes place in the root apex must obviously depend on adequate supplies of nutrients from the shoot, and on water and ions absorbed from the soil solution. Isolated root systems in culture, in addition to a balanced solution of electrolytes and a supply of energy-yielding carbohydrates, require certain extra micronutrients. These include such growth factors as indoleacetic acid (a growth-regulating substance) and thiamin (a vitamin of the B complex) which acts as co-factor to carboxylase enzymes concerned with the biochemical elimination of CO_2 during respiration. The growing root apex is thus by no means independent of the aboveground portions of the plant.

The pattern of cell growth at the apex of the root is somewhat variable in detail of a kind that need not be considered here. Reference to Fig. 7 · 1 (p. 99) will recall the essential features of the angiosperm root tip. Just behind the root cap is a *quiescent centre*. This is a zone of cells in which mitotic divisions are relatively infrequent. The evidence for this is partly anatomical, but it is also known from experiment (see Fig. 12 · 1) that the incorporation of tritiated (i.e.

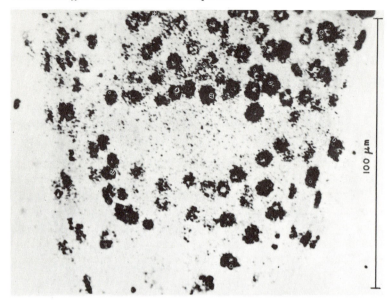

Fig 12·1 An autoradiograph of a longitudinal section of the root apex of *Sinapis alba* (white mustard), fed for three days with ³H-thymidine. The central clear area shows that no labelled thymidine has been incorporated into the cells of this quiescent centre during root growth. The surrounding cells have nuclei which have incorporated tritiated thymidine, as outlined in the text. From Clowes, *Apical Meristems*, by permission of the author and Blackwell Scientific Publications Ltd.

labelled with ³H) thymidine into DNA is minimal in this region, whereas in the surrounding mitotically active areas, DNA is being actively formed and can be detected by the effects on a photographic emulsion of the incorporated radio-active thymidine molecules (in which the tritium atoms present emit soft β-rays). The quiescent centre seems to act as a sort of template or shaping block, against which the main mitotic activity of the whole apex is organised. Cells produced by divisions at the proximal face of the quiescent centre contribute mainly to the tissues of the stele. Cells produced at the lateral flanks of the quiescent centre are primarily concerned with the later development of cortex and epidermis. The growing tip is covered by a root cap made of cells cut off to the front of the quiescent centre; these cells rapidly become vacuolated and by their mucilaginous breakdown help to reduce friction as the root forces its way between abrasive soil particles.

It should be remarked here that cells of the meristem possess varying numbers of very small vacuoles in their dense cytoplasm; by a joining-up and an extension of this vacuolar system older cells eventually come to show vacuolation that is visible under the light microscope. Characteristically, as each cell extends in length and girth, water is taken into the vacuolar system and the cytoplasm becomes almost entirely peripheral within the cell wall.

STELE. Vacuolation behind the quiescent centre occurs first in the stele, for the most part just after the earliest protophloem cells have declared themselves. Vacuolation starts centrally in the stele and spreads outwards toward the exarch protoxylem positions— nevertheless it is the protoxylem cells which complete their differentiation before the central cells in which vacuolation started. The first protophloem matures even earlier, and mature sieve tubes are therefore to be found nearer to the apex than mature protoxylem vessels. Protophloem and protoxylem occupy alternate radii at the periphery of the stele.

The earliest xylem elements stay slim, unlike the vacuolating central cells of the stele; they elongate rapidly and will later have to elongate much further. These vessel elements have to cope with the strains imposed upon them by the withdrawal of water from them into the shoot, and by the pressures upon them of neighbouring cells. They are strategically placed at the margin of the stele for the intake of water passing across the cortex and endodermal sheath, and they are mechanically supported by rings or by helices of lignocellulose in such a way that when extension growth of the root occurs, they can keep pace with it (at least at first) without loss of function. However, it is often possible at a later stage to see the remains of lignocellulose rings or helices from broken-down protoxylem left on the walls of cavities which have arisen through the vigorous stretching of the root; by this time the function of the protoxylem elements has been taken over by metaxylem vessels. As differentiation proceeds so cells mature closer and closer to the centre of the stele, though they already have been vacuolated for some time; this is referred to as centripetal differentiation. The central elements of the stele may differentiate as parenchyma, especially in monocotyledonous roots, but in many dicotyledonous roots the last-differentiated elements at the centre of the stele are often found to be well-lignified xylem elements.

The long, slim protophloem cells form strands which alternate in position with the protoxylem. They mature more quickly than the latter and they have densely staining contents and rather prominently shining water-imbibed cell walls. There is no further differentiation to give sieve tube and companion cell in protophloem, though this will happen to the phloem cells which differentiate centripetally next to them (p. 30 et seq.).

CORTEX. The cells of the cortex undergo most of their meristematic activity before phloem and xylem cells become apparent in the stele. Unlike these cells, which are elongated with respect to their diameter from a very early stage, cortical cells remain approximately iso-diametric, and only later, as elongation of the root as a whole sets in, does their longitudinal extension keep pace with that of the already extended stelar elements. Even so they remain essentially parenchymatous rather than prosenchymatous, and, since they are exposed to tensions during the process of extension growth, this is the point at which a fairly well developed system of airspaces begins to develop between the cells (see Fig. 4 · 8). Cortical tissue is a packing tissue and may also be used for storage purposes, but certainly one of its main functions must be to allow the ready diffusion of gases to and from the zones of intense metabolism and meristematic activity at the root apex.

EPIDERMIS. The origin of epidermal cells in the root is not always quite clear, for whilst their development may sometimes be traced from the quiescent centre, it is sometimes difficult to decide whether they do not share common origins with root cap cells. However, their maximum meristematic activity tends in any case to coincide with the first appearance of differentiated xylem elements, i.e. after the protophloem has appeared. Their subsequent extension growth coincides at its maximum with the maximum extension of protoxylem and protophloem, and as extension slows down, so one sees the lateral emergence of root hairs from the epidermal (or piliferous) layers, as seen in Fig. 7 · 2.

12 · 1 : 2 Stem apex

The working principles of the meristematic apex and of the processes of differentiation which go on behind it have been illustrated with respect to the root. No such analysis will be undertaken for the more complicated stem apex, though a very brief description will be given of a typical stem apex, in order that some of the points of contrast may be seen.

Figure 12 · 2 shows a series of horizontal sections through the apex of the common privet (*Ligustrum*). Its leaves are produced in opposite pairs, each pair alternating with those above and below them. Buds with this *decussate* leaf arrangement are most convenient to study, since a median longisection of the bud (see Fig. 12 · 3) traverses pairs of leaf initials, and one can therefore follow leaf differentiation in a leaf series in which succeeding leaf pairs are two *plastochrons* apart. (A plastochron is the interval of time that elapses between the appearance of one leaf primordium and its successor; in this instance pairs of leaves are involved.)

The apex forms a rather flat dome in *Ligustrum*; the dome may be higher in other plants. Leaf rudiments arise as small humps or papillae at the outer margin of the dome. There is controversy as to the control of their formation and appearance. Some say that they arise where space and nutritional factors are optimal. Others hold that they arise under intensely localised hormonal influences, probably the resultant of a balance between auxins from the dome itself and auxins from the already differentiating leaf primordia. Certainly the location of new leaf initials can be profoundly modified by the very local application of microgram quantities of auxin in lanolin (e.g. an opposite-leaved plant like privet can be temporarily caused to assume an alternate-leaved form, from which it ultimately reverts to its former leaf arrangement). Whatever the mechanism, there seems to exist a very fine degree of control in the location of leaf primordia with respect to their neighbours, and this is of course reflected in the arrangement of leaves in the adult shoot (known as *phyllotaxis*).

Once again, vacuolation plays a prominent part in the process of differentiation at the apex. In privet it is possible to discern a central zone of cells which vacuolate early on; these constitute the so-called *corpus*. Over these vacuolated cells lies a mantle of cells, varying in depth in different plants from two to five cells and known as the *tunica* (Fig. 12 · 4). Leaf primordia arise by local cell divisions which result in a heaping up of tunica cells towards the edge of the apical dome. The cells of the tunica divide and keep pace with the expansion (by vacuolation) of the corpus cells. Tunica cells divide mostly by cell walls laid down *anticlinally*, that is in a place at right angles to the surface. The origin of leaf initials can sometimes be detected by the *periclinal* divisions that precede their emission; in other words

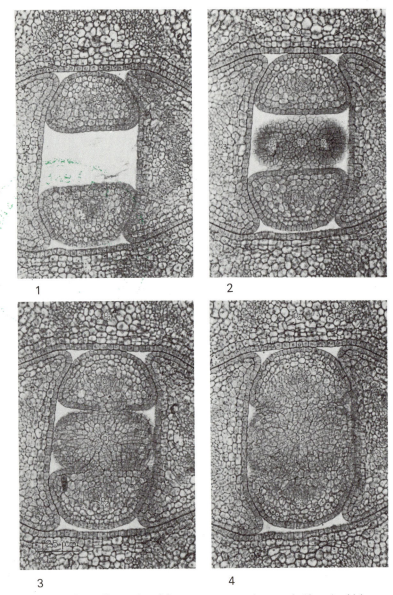

Fig 12·2 A descending series of four transverse sections, each 10μm in thickness, through the stem apex (ap) of *Ligustrum vulgare* (privet). The structure should be compared with the median longitudinal section seen in Fig 12·4. The leaf pairs are marked p₁, p₂, etc. in order of increasing age.

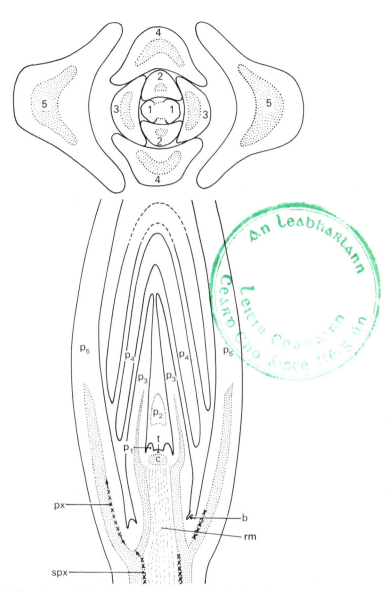

Fig 12·3 Drawings relating the transverse and median longitudinal views seen in Figs 12·2 and 12·4. Pairs of leaf primordia are numbered as before, the procambium is stippled, and the beginnings of protoxylem differentiation (px) are marked with crosses. It can be seen that differentiation in the leaf starts level with the leaf axil, and extends backwards to join the protoxylem of the stem (spx), and forwards into the extending leaf (see arrows). Phloem extends forwards from the stem and up onto the leaf, somewhat ahead of the xylem.

Abbreviations: t, tunica; c, corpus; rm, rib or file meristem; b, axillary bud.

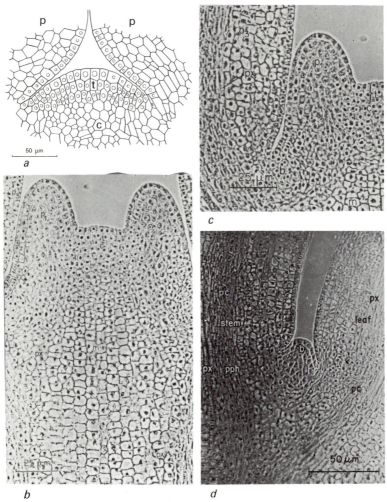

Fig 12·4a. Median longitudinal section (M.L.S.) of the apex of the stem of *Ligustrum vulgare*, showing the three-layered tunica (t) and subjacent corpus (c). The leaf primordia are well developed (p) and the apex is probably just at the beginning of a plastochron, with the dome at its highest. The new primordia are not in the plane of the section and cannot be seen.

b. A comparable axis seen at a later stage, the leaf primordia (p) being further developed and in the plane of the section. The two-layered tunica (t), corpus (c) and rib meristem (m) can be seen. At two positions it is possible to see evidence of differentiating tracheids.

c. A wider view of the same section shows protoxylem just beginning to form in leaf p_3, and the early beginnings of bud formation (b).

d. L.S. at a lower node, probably p_5, where a bud has developed in the leaf axil. Stem protoxylem is well established inside the procambial strand (pc) in the exarch position, and also adaxially in the leaf. [Ph.]

152

cell walls are laid down more nearly parallel to a tangent to the domed surface, and herald the building up of the minute hump.

The process of vascular differentiation is seen to differ in detail from that in a root, and is closely linked to leaf emission. Each leaf papilla develops a buttress where its base juts out from the apical dome; the buttress is at first surmounted by a cylindrical papilla growing in the early stages almost entirely by means of an apical meristem. Very soon marginal meristems are defined which anticipate the appearance of the flange-like outgrowths that eventually form the leaf lamina. Vascular strands begin to differentiate and protophloem grows forward from leaf traces in the young stem below, and up into the emergent leaf. On the other hand the differentiation of xylem starts at the level of the axil of the young leaf (i.e. where its upper surface joins on to the stem and where one may often see at a very early stage the beginnings of a bud). Differentiation of xylem extends forward into the developing leaf and backwards to join on to the existing vascular system of the older stem and leaves below. This may be related to the Skoog experiment already quoted (p. 141) and suggests that vascular differentiation is induced hormonally. Thus it is seen that stem anatomy is a composite of contributions from the leaves which are produced in succession at the apex; in other words the anatomical framework of the stem is largely composed of leaf traces.

In experiments where leaf rudiments are carefully excised, it is possible to show that stems will still become vascularised, albeit in primitive fashion since they produce a solid rod or an unbroken cylinder of vascular tissue akin to the structures seen in the more primitive ferns. As the apex returns to normal after such an operation, leaf initials are produced according to the normal pattern, and so also the vascular structure of the stem resumes its customary dissected form.

The stem apex does not only produce leaf rudiments, of course, and it is important to recognise that bud rudiments arise very early on in the axils of the young leaves. These bud rudiments though formed early may not develop appreciably until much later, and even then, as described in section 13 · 5, may be restrained in their growth by the action of inhibitors. The stem apex itself is profoundly affected, as will be seen in the following section, by factors which induce in it considerable changes of behaviour, as it passes over into a reproductive phase.

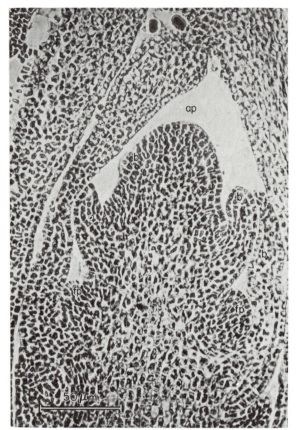

Fig 12·5 L.S. of the reproductive apex (ap) of *Salvia pratensis* (meadow sage) which has a considerably higher dome than in the vegetative state. Bracts (b) subtending flower primordia (fp) can be seen.

12 · 2 *The contrast between vegetative and reproductive growth at the stem apex*

In Part 4, which deals with the physiology of flowering, we shall learn that a vegetative plant eventually reaches a stage which has been termed 'ripeness-to-flower'. At this stage changes take place in the stem growing point which herald the onset of flower production. The steady pattern of leaf rudiment formation gives place to a phase in which each apical meristem changes its mode of growth

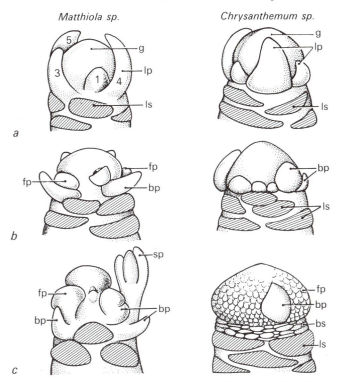

Fig 12·6 Stages in the change from vegetative to reproductive growth in *Matthiola sp.* (Virginian stock) and *Chrysanthemum sp.*

a. Vegetative growth; only leaf primordia (lp) are present, and where leaves have been cut away to expose the growing point (g), leaf scars are seen (ls).

b. Leaf primordia give place to bract primordia (bp). In *Chrysanthemum sp.* these later form a protective involucre.

c. The flower primordia of stocks (fp) develop in the axils of bract primordia, and in the oldest of them sepal primordia have already developed (sp). The involucral primordia of chrysanthemums are succeeded by a series of floral primordia, development of which gives rise to the characteristic composite inflorescence, the capitulum.

in such a way that it can accommodate more primordia at one time. In so doing it generally becomes wider and often much more obviously domed (see Fig. 12·5). The grouping of flowers on a stem axis (known as the *inflorescence*) is also apparent at this stage, and unless the inflorescence is restricted to a solitary flower, the apex may become more or less crowded with a sequence of leaf and bud primordia (see Fig. 12·6). Inflorescence patterns fall into two main categories, though the distinction between these categories is

155

not always clear cut. In the first category the growing point is used up in the production of a flower, and, if that flower is not solitary, all subsequent flowers arise from buds developed in the axils of leaves (known as bracts), so that one can generally see two kinds of rudiment in a flowering apex. This gives rise to what is known as a *cymose inflorescence*. In the other category of inflorescence pattern, a *racemose* inflorescence is formed in which the growing point stays active, i.e. it is never used up in flower-bud formation, though it continuously gives rise to bud rudiments below it. In a raceme such as the flowering spike of lupin, the oldest flowers are at the base of the spike; flowers are progressively younger towards the apex of the spike, and growth is termed indefinite since the apex continues to produce new groups of primordia. Thus the collections of primordia which constitute a bud rudiment mostly, but not always, arise in the axil of a bract rudiment; all of these rudiments are laid down at the persistent growing apex.

So that each individual flower rudiment develops as if it were an axial (stem) structure of limited growth, bearing leaflike structures (sepals and petals) developed from papillae similar to vegetative leaf rudiments. These are succeeded by primordia which grow into stamens (sometimes showing a measure of leafiness) and then lastly by the ovary. Theory has tended to equate the ovary with a fused multiple fertile-leaf structure, but this is a controversial issue which need not concern us here. What really matters is to understand that, in both stamen and ovary, two new events take place (*meiosis* and *sporogenesis*), as a result of which haploid gametes are formed, and later fuse; the new diploid organisms which arise from this fusion show a freshly shuffled assortment of hereditary characters.

The physiology of flowering, then, is concerned firstly with the factors that induce these changes in the vegetative growing point, and secondly with the timing and control of all the subsequent events that lead in the end to the laying down of the seeds.

13 Hormonal control of growth in plants

13 · 1 *Introduction*

A hormone is often described as a chemical substance, normally effective at low concentration, that is produced at one locus and may then bring about its effects elsewhere. There are four main groups of plant hormones, and these include auxins, gibberellins, cytokinins and abscisins. There is little doubt that substances in groups other than these will be disclosed from time to time and will also be shown to contribute to one aspect or another of the hormonal control of growth.

Plant hormones appear to be manufactured mostly at centres of high anabolic activity, i.e. in meristematic and rapidly expanding tissues, and the balance between these four groups of substances has a profound effect upon the manner of cellular growth. It has been mentioned already that the growth of undifferentiated callus tissue in *in vitro* culture can be more or less modified by varying the proportions of auxins and kinetin present. It seems as though these substances between them can cause cells to undergo mitosis (i.e. either they remain actively meristematic, or, if they are not already active in this way they may revert to the meristematic state and to active synthesis). Alternatively, cells may move towards a phase of expansion (largely longitudinal) and differentiation, as discussed in Chapter 12.

It now seems very probable that auxins and gibberellins may help to bring this differentiation about by action at the level of the gene, viz. by altering the type of messenger RNA synthesis, or in other words by changing the programming for protein and enzyme production and thus by changing the behaviour pattern of the cell. So, too, the cytokinins are believed to be involved chemically in the activity of transfer RNA, which is not surprising since some cytokinins are also, like t-RNA, derivatives of purine, e.g. adenine.

157

Certain limited aspects of these relationships have already been discussed in *Molecules and Cells* in this series.

In a book which aims to view the plant rather than the cell as a working unit, these important molecular aspects must nevertheless give place to more broadly physiological views. Before doing so, however, it may be useful to take a very much restricted look at the chemical nature of these substances.

The *auxins* of plants are for the most related to or based upon the substance known as indole, a double ring structure containing nitrogen. The chief representative of this group of substances is indoleacetic acid (IAA).

Numerous other compounds have been synthesised which have auxin-like properties. Some (like 2,4-dichlorophenoxyacetic acid, or 2,4-D) are known better to farmers and horticulturists as herbicides; however, there begins to be wider recognition of the dangers to wildlife of these active compounds, which by incorporation into food chains may even lead so far as to diminish the breeding potential of the tertiary consumers (carnivores). Homologues of IAA (i.e. compounds with slightly different though basically similar structure) have been synthesised in abundance, and, apart from the herbicides, a number of these homologues are used in horticulture. They are used for a range of purposes, including the rooting of cuttings and the control of premature abscission and fruit drop in commercial orchards.

The *gibberellins* have been known to plant physiologists for a much shorter time. Chemically they belong to the group of compounds known as the diterpenoids, and their molecules have a five-ringed structure (see below) that is too complex to discuss here. Some of them may conveniently be isolated from fungal extracts; indeed their effects were first studied in rice plants in which abnormal extension of the inflorescence had occurred as the result of attack by the fungus *Gibberella*. Upwards of 30 related compounds have now been isolated but the most active is gibberellic acid (GA_3):

$$
\begin{array}{c}
\text{H} \quad \text{O} \qquad\qquad \text{H}_2 \\
\text{C} \qquad\qquad \text{H} \quad \text{C} \\
\text{HC} \quad\quad \text{C} \text{---} \text{C} \qquad \text{CH}_2 \\
\text{| CO} \quad\quad \text{|} \qquad\qquad \text{|} \\
\text{OHHC} \quad \text{C} \qquad\quad \text{C} \qquad \text{CHOH} \\
\text{C} \quad \text{H} \quad \text{C} \qquad \text{C} \\
\text{|} \qquad\quad \text{H}\backslash \text{-----} \text{C}=\text{CH}_2 \\
\text{CH}_3 \qquad\quad \text{COOH}
\end{array}
$$

The *cytokinins* are a mixed group of substances, in which the most interesting are perhaps those based upon the molecule of adenine:

$$
\begin{array}{c}
\text{NH.R} \\
\text{|} \\
\text{C} \\
\text{N} \qquad \text{C} \text{---} \text{N} \\
\text{|} \qquad\quad \text{||} \qquad\quad \text{CH} \\
\text{HC} \qquad \text{C} \text{---} \text{N} \\
\text{N} \qquad\qquad \text{H}
\end{array}
$$

Thus this group is already related to a wider range of important substances in the plant (including ATP, RNA, DNA etc.) and it is hardly surprising that they are so active in growth.

Abscisic acid (or ABA) is a comparatively recent discovery and is a very good example of a hormone that primarily acts as an inhibitor of growth. It has now been synthesised and shown to have the formula:

$$
\begin{array}{c}
\qquad\qquad\qquad \text{CH}_3 \\
\qquad\qquad\qquad \text{|} \\
\text{H}_3\text{C} \quad \text{CH}_3 \quad \text{H} \\
\text{C} \qquad \text{C} \text{=} \text{C} \quad \text{C} \text{=} \text{C} \quad \text{H} \\
\text{H}_2\text{C} \quad \text{C} \qquad\qquad\qquad \text{COOH} \\
\text{|} \qquad\quad \text{|} \quad \text{OH} \\
\text{C} \qquad \text{CH} \\
\text{O} \qquad \text{C} \\
\qquad\quad \text{H}
\end{array}
$$

There is some evidence that compounds similar to ABA may occur in plants (e.g. abscisin I in cotton plants and phaseic acid in seeds of *Phaseolus* the runner bean) and it is almost inevitable that with increased research pressure upon this group of substances we shall come to learn more of its variety. So far as we at present are aware, ABA itself is the most potent of them all.

13 · 2 *Auxins*

13 · 2 : 1 Action and mode of action of auxins

Using IAA as representative of the auxins, we may note that it is normally produced in buds (i.e. in meristematic apices and the active young leaves that surround them), as also in actively growing embryos. Work on leaves has indicated that although the *intensity* of auxin production is greatest in young unexpanded leaves, quantitatively leaves produce most auxin during their period of maximum expansion. Auxins in the shoot are transported in a strictly basipetal direction, viz. from shoot to root and they continue to move in the roots towards the root apices. This basipetal stem transport occurs even when the stem is turned through 180°, so that it cannot be ascribed to gravity, and must depend on some kind of polarisation of the cells in which it is travelling. Auxin transport seems to follow most of the principles established for many organic materials (see section 8 · 3) but the polarised direction of its transport distinguishes it from the movement through the plant of sugars and similar organic molecules.

Auxins, apart from the part that they play in stimulating mitotic activity, appear to promote extension growth in stem tissues at an optimal concentration of the order of ten parts per million. It is not yet known with certainty how extension growth is achieved but it certainly involves (*a*) a loosening of the microfibrils of the primary cell wall; and (*b*) the adjustment of turgor pressure to make possible an expansion of the cell, generally in the longitudinal direction.

Some protein synthesis is stimulated at the same time, so that an increase in cytoplasmic materials goes hand in hand with the extension process. It is not yet fully established that new cellulose materials are simultaneously synthesised during extension, though there does seem to be good evidence of an increase in the amount of pectic materials in the wall. However, cellulose materials are added soon afterwards to stabilise the loose microfibril framework which has undergone extension.

13 · 2 : 2 Extraction and bio-assay of auxins

Indole auxins may be detected and assayed by various chemical means, but the techniques are hardly sensitive enough for the levels usually encountered and it is customary to employ so-called bioassay methods. The chief of these methods involve either direct measurements of extension growth, or measurements which depend on the differential growth curvature that results when auxin is applied to one side of an erect stem or coleoptile. It is proposed here to deal at rather greater length with one of these techniques, so that those interested can explore some of the activities of growth regulators for themselves.

Coleoptiles have been used extensively as sensitive test objects for the assay of growth regulators. A coleoptile is a hollow sheathing organ that surrounds the growing shoot of grasses (including of course the cereals); by its own extension growth the coleoptile helps to penetrate the soil, and the protected shoot bursts through the coleoptile sheath when the latter has reached the end of its growth. (Fig. 13 · 1a). During this period of extension growth, auxin is produced at the coleoptile top and, passing back basipetally in the customary manner, stimulates cell elongation along its route.

Normally oat (*Avena*) coleoptiles are used for auxin testing, though for most class purposes wheat coleoptiles serve nearly as well, and are not so tricky to handle. For instance, certain varieties of oats require dehusking before use, and subsequently require a brief exposure to red light in order to suppress growth of the mesocotyl, viz. the first internode above the grain. Wheat grains should be soaked for a preliminary 2–3 hour period, and then planted out groove downwards on to two or three thicknesses of wet tissue or filter paper on a dish in a light-tight box (Fig. 13 · 1b). The dish is placed at an angle of about 45° with the horizontal so that the vertically growing coleoptile grows away from the sloping grain; the box should be placed in a warm cupboard. The growing time should be from three to five days according to the temperature.

STRAIGHT GROWTH TESTS. For assay by measurement of extension growth, coleoptiles are removed from the germinated grains, and a special cutter and block (Fig. 13 · 1c), with razor blades appropriately spaced between pieces of Perspex, may be used to remove the apical 5 mm of the coleoptile, whilst retaining the next 1 cm segment below it. (Similar segments may be cut from roots or

161

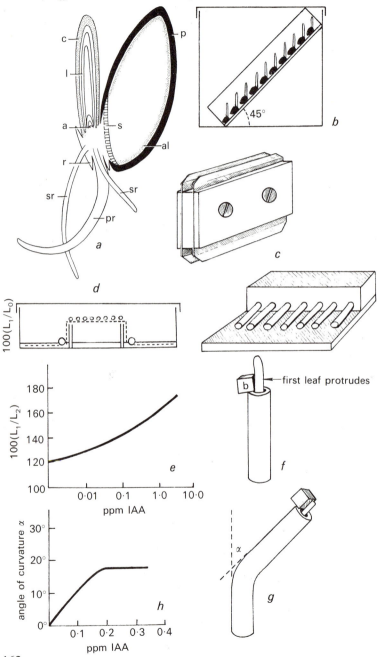

from the first internode below the plumule in dark-grown etiolated pea seedlings.) The 1-cm segments are accumulated on to damp filter paper as they are cut, and subsequently transferred to petri dishes containing the test solution or IAA at the appropriate concentration, together with about 1% of sucrose as an energy source. Aeration is important, and the preparations are sometimes agitated in a small volume of fluid during the prescribed twenty-four hours of growth. Alternatively and perhaps most conveniently they may be allowed to grow on the surface of filter paper, wetted by capillarity by the medium in use (Fig. 13 · 1d). With a sucrose-containing medium, normal precautions should be taken to reduce infection, and solutions should not be kept longer than five to six days after making them up.

A number of *extraction procedures* have been used for obtaining auxins from living tissues. The simplest technique is probably the best, namely to allow the auxins to diffuse into a minimal amount of water, or into 1% agar gel. If chromatography is possible, such an aqueous extract can be separated out into its growth-promoting and growth-inhibiting components; the dried chromatogram is divided into strips and each strip can be assayed by the method of Fig. 13 · 1d. The use of other solvents may result in the extraction

---

Fig 13·1 The use of coleoptile segments in the bioassay of auxins.

a. L.S. through germinating grain of *Avena sp* (oat), showing the coleoptile (c) and its enclosed shoot, consisting of the first leaf (l) and the apex (a), and also the ruptured root sheath or coleorhiza (r), through which has grown the primary root (pr). Note also seminal roots (sr), endosperm (e), scutellum (s), aleurone layer (al) and pericarp (p).

b. A method of growing coleoptiles in a dark box.

c. The mounted razor blade assembly and perspex block used for guillotining coleoptile (or root) segments.

d. The arrangement for growing the segments on a filter paper strip wetted by the test solution.

e. A graph showing the relationship between percentage increase in segment length and auxin concentration.

f. An arrangement for measuring the bending response. A 5 mm segment of the tip has been removed from the coleoptile, leaving the first leaf protruding. An agar block (b) 5–10 mm³ in volume, and cut from 3% agar sheet containing auxin or the substance under test, is placed eccentrically on the coleoptile stump, supported by the pulled-out leaf.

g. Shadowgraphs of the bent coleoptiles are prepared and the angle of bending is measured as shown.

h. A graph showing the relationship between the bending response (curvature) and auxin concentration.

of bound and inactive forms of auxin. The use of water as an extractant at least makes it possible to assay free active auxin, diffusing in its normal medium.

CURVATURE TESTS are more demanding in skill and in the control of growth conditions. They depend on the placing of a small block of agar containing the growth substance under test to one side of a decapitated coleoptile (about 2·5 cm in length), as in Fig. 13 · 1f. The mean angle of curvature is used, with suitable statistical controls, as a measure of the auxin concentration in the block. The curved coleoptiles are generally shadow-printed on to a sensitive photographic paper, and measurements are made on the images so obtained (Fig. 13 · 1g). This test is quite sensitive over a range of concentration from 0·001 parts to about 0·2 parts per million (2×10^{-7}). The straight growth test shows smaller sensitivity, but shows it over a greater range, from about one part in a thousand million (10^{-9}) up to about one part in 100 000 (i.e. 10^{-5}) (see graphs Fig. 13 · 1e and h).

13 · 2 : 3 Growth curvature and auxins
It has long been known that coleoptiles show a growth curvature in response to exposure to light from one side. We have already seen that the deliberate one-sided positioning of an auxin block on a decapitated coleoptile brings about a response in curvature, whose measurement can be used in assaying auxin concentration. Suggestive as this is about the possible effects of an uneven distribution of auxins, it is by no means yet confirmed that the bending response of stems and roots is specifically concerned with the differential distribution of auxin in their growing tissues.

Where the differential response in growth of an organ is due to light, we talk about a *phototropic* response. A tropic response is a *directional* growth response of an organ to a *directional* stimulus, in this case light. A *nastic* response, by contrast, is a response by growth or turgor change in relation to a diffuse, *non-directional* stimulus, such as a change in temperature, as for example when flower petals open on bringing flowers into a warm room.

We know that under the influence of unidirectional light, the concentration of auxin which reaches the zone of extension growth in a stem becomes lower on the illuminated side and higher on the darkened side. This has been attributed by some, partly to the destruction or inactivation of the IAA by light at the blue end of the spectrum, and partly to a lateral transport of auxin away from the lighted side. The tissues on the darkened side show a larger growth response than those on the lighted side, and a positive growth curvature results, i.e. the stem bends towards the source of light. In stems and roots, gravitational forces can be shown to produce a *geotropic* response, though the mechanism is still far from being understood (even after more than 80 years of study dating back to Darwin's first observations on movement in plants). We still do not know why stems and roots show a different geotropic response, the stem being negatively and the root positively geotropic. It may well be shown eventually that gravity affects the growth response of roots and shoots by controlling the distribution of *inhibitors* of growth substances, rather than the growth substances themselves. A further possibility is that gravity affects the rates of transport of the growth regulators concerned, but the problem still remains to be satisfactorily solved as to how the plant perceives the stimulus of gravity. In the root it has long been suspected that the starch grains (and maybe plastids) in the root cap act as the means of stimulus perception, by falling through the cell on to the lowermost cell wall; the mechanism for perception is not yet clear.

Curvature by differential growth is seen on a number of occasions in plants; the turning of flowers to face the sun, the erection of a drooping poppy bud as it flowers and starts to develop its capsule, the twining of tendrils, petioles and stems, all of these involve differential growth extension on opposite sides of the organ concerned as a result of the differential concentrations of auxins present.

13 · 3 *Gibberellins*

Substances in this group are also essentially concerned with extension growth; whilst they generally work in association with the indole auxins, they are quite capable of independent action. Gibberellins, like auxins, appear to be produced by cells with a high rate of synthesis and metabolic turnover, viz. in young and expanding tissues. Their transport in the plant, unlike that of the auxins, is not polarised, and whilst in normal growth both groups are concerned with cell elongation, only gibberellins can stimulate extension growth in the stem axis of certain (mostly biennial) rosette plants, and of dwarf mutants (in which presumably the genetic factors for normal gibberellin production are absent).

The apical meristem in such plants is active in the normal way but it is the *sub-apical* region of these meristems in which activity depends on the availability of gibberellins; they appear to stimulate further cell divisions in which the new walls are mostly laid down in a transverse plane, at right-angles to the long axis of the stem, as in the so-called rib-meristems. Cell division is followed by extension growth in which auxins play a normal part. There is no comparable sub-apical region in a root, and the gibberellins appear to have little effect upon the growth of roots in general. However, there is some evidence that root apices may produce gibberellins, the presence of which has been demonstrated in root exudates, i.e. the sap which exudes from the cut end of a decapitated root or root system.

Dwarf mutants of *Zea mais*, Meteor peas and rice seedlings have all been used as test objects in gibberellin assay, but the effects of gibberellin stimulation can be easily demonstrated and the increase in growth that turns a bushy pea or bean plant into a scrambling climber is most striking. A drop of solution containing as little as one tenth of a microgram (0·1 μg) of GA_3, placed in the axil of a young leaf, will cause a dwarf pea plant to double its height in under three weeks.

(A typical bio-assay procedure is to soak Meteor or Alaska peas for six to eight hours, and to germinate them in moist vermiculite for four days at 20°C; they can then be stood in the aqueous extract under test, under white light. The length of the epicotyl is measured after a further five days' growth, and the test is claimed to be sensitive to the equivalent of 1 ng or 10^{-9} g of GA_3.)

The gibberellins are most clearly implicated in the balance between dormancy and growth, and can be used as a means of breaking dormancy in buds and seeds. The onset of dormancy seems to be accompanied by the lowering of gibberellin levels and the increase of growth inhibitors such as abscisic acid (ABA). However, once seeds have been germinated, the endogenous production of gibberellins by the seedlings gradually rises. It is known too, that gibberellins stimulate the synthesis of enzymes that hydrolyse starch and proteins. Thus the break of dormancy is accompanied very appropriately by an increase in the availability of sugar and amino-acids for subsequent growth.

Gibberellins are thought to be concerned also with the change from vegetative to reproductive growth at the apex (see also the following section), and it is worth drawing attention here to the fact that the effects described so far in all likelihood result from the multiple presence of different kinds of GA molecule. Most bio-assays for this group focus attention on growth extension, but we may be soon unravelling the effects of gibberellins not specially active in vegetative growth, but most certainly of comparable importance in other spheres of plant activity, and particularly in reproduction.

13 · 4 *Cytokinins*

Reference has already been made (section 11 · 1) to the balance that exists between auxins and cytokinins in controlling the growth behaviour of cultures of plant tissues, as well as to the fact that the cytokinins in the broadest sense embrace a rather diverse group of chemical compounds. In discussing the various roles of the cytokinins we shall have in mind the purine derivatives in particular, for these are in any case the most active of the cytokinins so far identified. Kinetin itself has not yet been isolated as such from plants, though this does not detract from its value as a highly active synthetic compound which can also be recovered from autoclaved DNA from various plant and animal sources.

It is thought that the main role of the auxin + cytokinin system lies in the regulation of mitotic activity. It will readily be seen that there is a need for nuclear control, not only in the meristems of root and stem, but also in connection with the onset or release of mitotic activity in previously dormant buds, embryos and secondary meristems. The auxins seem to be able to promote in cells the doubling up of chromosomal DNA during the inter-mitotic phase. Cytokinins, on the other hand, have been shown to increase RNA synthesis and thus an increase in protein metabolism. This may help to explain why, although IAA occasionally triggers off mitosis, this usually only results in binucleate cells; on the other hand cytokinins stimulate both mitosis and cytokinesis, and the process of nuclear division is completed by cell division involving the laying down of phragmoplast and cell wall (p. 3). In addition to stimulat-ing embryo (and embryoid) growth in tissue cultures (see p. 141), the synthetic kinetin can evoke bud formation in callus masses; its relationship to IAA in this respect is shown in Fig. 11 · 1.

Cytokinins play an important part in the expansion and differentiation of leaf tissues, and presumably, therefore, in the morphogenesis of whole leaves. Furthermore kinetin has been shown to stimulate a general increase in synthetic activity and metabolic turnover in leaf cells. In a leaf that has been treated locally by the application of a drop of kinetin solution it can be shown that reserves may be mobilised from some distance away from the drop, and that sugars and amino-acids move towards and collect in the very localised area affected by the drop. This seems to suggest that kinetin itself may not be very mobile under these circumstances. Nevertheless, there is evidence that naturally-occurring cytokinins may be detected in the exudate from roots, suggesting that these substances are manufactured at root apices, and may be moved up into the shoot system, where they augment those already available in the shoot.

After painting a kinetin solution all over the surface of a detached leaf, it is also possible to demonstrate the arrest of the processes of ageing and in particular of chlorophyll degradation that would normally be expected to occur. It is comparatively easy to assay chlorophyll in a leaf by extraction with boiling alcohol (do not use a naked flame for this). One can subsequently measure the optical density of the extract in a colorimeter, after making up to standard volume. Denser chlorophyll solutions will be recovered from detached leaves that have been treated with kinetin and left for two days than from comparable leaves just wetted with water.

The arrest of senescence in leaves that is demonstrated by the chlorophyll test is not beyond reproach as a bio-assay method. A number of substances that are not cytokinins are believed to be intimately concerned with the RNA activity which mediates protein synthesis. Newly-stimulated protein synthesis may provide structural units for new cytoplasm, or it may provide enzymes for increased metabolism; and presumably the manufacture of chlorophyll in the leaf involves both kinds of protein. One may perhaps think of a system such as this in three phases:

1. a highly active phase in which protein manufacture and turnover is such that a 'metabolic sink' is created, to which raw materials (amino-acids etc.) move from regions of higher concentration,

2. a less active phase in which protein turnover is taking place at a lower 'maintenance' level, not necessarily yet involving senescence, and

3. a reduced level of protein synthesis that is inadequate to stop the system from running down in the process of senescence.

Cytokinin-stimulated activity, corresponding to phase 1 above, may be held to be considerably involved in the expansion of cells, particularly during the growth of leaves, as well as for much of the activity in cell division which precedes this expansion stage.

13 · 5 *Growth hormones: interactions and inhibitions*

We need first to remind ourselves that it is only in the last two decades that plant physiologists have ceased to think of growth so single-mindedly in terms of auxin effects. We are aware nowadays of many more factors that enter into the processes of growth of cells and tissues, and it is very probable that more growth factors may yet be disclosed to add to the complexities of the present situation. Not only are we now concerned with the balance of known hormones, in terms of their presence or absence or their concentration effects, but we are also concerned with the intervention of other substances which may oppose their activity. These substances, as it were, interpose blocks or barriers between the regulators and the cell processes that are being regulated.

In the routine assay of tissues for growth regulating substances, it is customary nowadays to make an aqueous extract by diffusion from the tissue concerned; this extract is then concentrated and chromatographed. The chromatogram is divided into equal portions along its length, and the contained substances are eluted from each individual portion and tested for their efficacy by means of an appropriate assay method. Figure 13 · 2 gives an example of the kind of information that can be obtained in this fashion. It shows that there is an active growth-influencing substance present, and also a substance which retards growth in this test.

It is possible that molecules which show such growth inhibition compete with auxin molecules because they have certain similarities of shape, and can thus fit on to the substrate normally occupied by auxin (though in fact they lack the growth activity of the true auxin). They are competing for available auxin sites. Other growth inhibitors may interfere with auxin action by interfering with auxin transport to the site of action. Growth may be strongly inhibited by yet another group of substances which appear to act by antagonising gibberellin activity. For example, the effects of the commercially produced

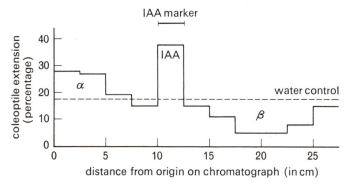

Fig 13·2 A histogram showing the results of the bioassay of a chromatographed extract from etiolated bean shoots. Each 2·5 cm portion of the chromatogram is tested for its effect on the growth of standard oat coleoptile segments. A spot of IAA has been allowed to run on the paper as a marker, and corresponds in position to the peak in growth activity shown at 10·0 cm to 12·5 cm.

synthetic growth inhibitor known as CCC can be offset by gibberellin treatments, and this is equally true of its inhibitory effects upon seed germination and upon the onset of flowering in long-day plants (see next section).

It will be grasped that most aspects of the growth and differentiation of plants are under multi-hormonal control, and that reaction, interaction and inhibition may all contribute something to the processes whereby shoots and roots become fully differentiated as mature organs.

The growth of an organ involves (as already stated in Chapter 11) the differential development of specific types of cells in specific tissues, with all their inherent points of contrast with the nearby cells of other tissues. But the differentiation of an organ, involving the processes of morphogenesis, takes account of more than the maturing of groups of individual cells. Thus in the early stages of organo-genesis it is not just cell division which is implicated; the shaping of an organ depends also on the locally varying rates of mitosis within that organ, on the plane of the cell divisions themselves and therefore on the orientation of the cell plates. We shall do well if we ever manage to solve the problem of how these phenomena are controlled in order, say, to account for the differences between a pine needle and a holly leaf.

During the development and growth to maturity of an organ, two main processes are involved. The first process, cytokinesis, helps to shape organs by variations in its rate, location and orientation within the organ. For example, the application of GA to the stem of a rosette plant induces cell division in the subapical meristem; by these divisions cell walls are predominantly laid down in a plane at right angles to the long axis of the stem. When expansion growth occurs, this results in a striking extension of the stem axis. Here the effect of the mainly transverse planes of cell division is reinforced by the longitudinally polarised extension growth of the cells produced.

However, the second process (i.e. expansion growth) is not always longitudinally polarised; one has only to think of the way in which a pumpkin or an apple expands to realise this. Again, in the typical dorsiventral leaf, a whole range and sequence of cell behaviour is seen. Vascular strands develop mostly as they do in the stem, i.e. tracts of procambial tissue increase in girth by cell divisions with cell plates in longitudinal planes. The laying down of transverse (horizontal) walls combined with longitudinally polarised extension growth leads to an extension of the whole strand system, whether the venation is reticulate or parallel (p. 59). Epidermal cells expand more or less equally in all directions in the plane of the leaf, and except in monocotyledonous leaves show polarised extension growth only in the region of the veins. Palisade cells grow at first like the epidermal cells but gradually show extension growth at right angles to the plane of the leaf. Both they and the epidermal cells at first divide by anticlinal walls, typically providing an extending skin of cells over the vascular framework. Then, as explained on p. 67, the rate of cytokinesis slows down in the palisade cells, but continues steadily for a while longer in the epidermal layers before slowing to a stop. Epidermal expansion takes place in the plane of the leaf, and the subsequent behaviour of these and the mesophyll cells has already been described ($5 \cdot 2 : 2$). What matters more in the present discussion is that we recognise that there is a subtle time-tabling of the events described. A behaviour pattern is established somehow for epidermal cells, which is different for the palisade cells immediately next to them, and different again for the fundamental vascular tissues which supply their needs and drain away their products.

171

As we know, cells form tissues, and tissues are constituted into organs. The form of the whole plant is determined by the disposition of the organs which are part of it, and even these organs are subjected to the overall form of control which is known as *growth correlation*. There can be few who are not broadly aware of the difference in form, say, between a spruce and an oak tree, yet it is important to recognise that total plant form is in itself an expression of the integrated interactions of growth regulators and growth inhibitors.

Growth correlation in general depends in considerable measure upon the control of localised apical growth, whether in lateral buds or in lateral roots. This control is closely associated with the activity of the terminal bud or leader, and the over-ruling control exerted by the leader is known as *apical dominance*. Some simple experiments with pea seedlings may be used to illustrate this process. The growth of the main shoot in pea seedlings suppresses the development of the buds in the axil of each cotyledon. If the main shoot is decapitated (i.e. the leader is removed) both of the cotyledonary buds start into growth. If the terminal bud is left intact, whilst one cotyledon with part of the main stem and root system is isolated from it, the bud in the axil of the separated cotyledon will grow out. If, however, the severed unit is left in contact with the remainder of the plant, being separated from it only by a film of water, something diffuses across the aqueous gap which arrests the growth of the cotyledonary bud. If IAA is applied in an auxin block to a decapitated main axis stump, axillary bud growth is still inhibited. It has been established that inhibited buds can be released from inhibition by the direct application to them of kinetin, though apical dominance may supervene if the treatment is not continued for such a time as to allow the extending shoot to manufacture enough of its own auxin.

Space does not allow any detailed discussion here of the *mechanisms* of correlative inhibition. The presence of auxins at the apex may be instrumental in diverting the translocation of metabolites from the dormant lateral buds. Bud dormancy has in some cases been shown to be partly due to the incompleteness of the vascular connection between the bud and the main vascular supply of the stem; with the removal of the auxin supply from the dominant apex, vascularisation may proceed rapidly and may thus allow a less restricted supply of nutrients to the arrested lateral bud.

Whatever the controls that may be operating, the form of a plant generally reflects at least something of the degree of dominance of the apical bud. A climber or straggler (such as *Clematis* or *Cucurbita*) shows a suppression of lateral branching, and axillary buds develop mostly in response to the stimulus to flowering that eventually arises. Some woody plants such as apple or larch characteristically show *dwarf shoots* of limited growth, which develop from lateral buds on *long shoots*. These lateral buds are only partially suppressed and produce their yearly quota of leaves and perhaps flowers, but they contribute nothing to the extension of the branch. Larger woody plants are often less easy to explain in terms of simple dominance. In our original contrast between spruce and oak, the situation for apical dominance is in reality rather different from what might be expected. On the current year's growth in oak, the lateral buds that are laid down are firmly suppressed, and are only released from dominance after a winter of dormancy. Their control seems to start with apical dominance by the leader, and to be continued by the accumulation within the bud of inhibitory substances (not auxins); the buds require a cold exposure before growth release is possible and they can be released by a variety of other treatments. Thus in oak which has the spreading crown of branches typical of sympodial growth (where the leader fails and the laterals take over), nevertheless there exists this pattern of local apical dominance. In the case of spruce, where growth is normally sustained by a dominant leader (monopodial growth) when we come to look at the growing point we find a number of laterals growing just behind the leader; it could hardly be said that apical dominance is very marked in this tree. Nevertheless it is only if the leader becomes damaged, that one of the more vigorous laterals quickly takes over as leader. The pyramidal form, so typical of some conifers, is of course not merely dependent upon monopodial growth in which height increase is mediated through an established leader. Uppermost branches grow out at an angle to the main stem. Lower branches, though at first making this angle with the vertical, later (and then largely because of their increasing weight) come to take up a descending angle with the vertical.

It will readily be seen, too, that the form of trees and shrubs may also be partly determined, not only by the angle that the branches make with the main stem, but by such factors as whether the lateral buds that develop are on the physiologically upper side or lower side of the branch. It is hardly surprising that the form of trees, shrubs and herbs can show so much diversity in view of all the possibilities for modification and control that exist.

Mention has been made above of the direction of growth of laterals with respect to the main shoot. This is the phenomenon of *plagiotropism*, and it can affect such diverse organs as roots and underground stems, as well as the leaf petioles and branches of the above-ground system. Little enough is known of the control mechanism which leads to plagiogeotropic response in roots. In plants such as the lupin, where the primary root acts as a deeply penetrating taproot, control of the growth of lateral roots does not merely keep in check those laterals which are nearest to the main apex, but somehow induces them to grow away from the taproot at a descending angle to the vertical. This is clearly advantageous to the plant if only in terms of the improved exploitation of the soil volume around the taproot. In fibrous root systems, for example in the annual grasses such as wheat and barley, there seems to be little correlation between the growth of the main members of the system, though each individual member may produce laterals over which its growing point appears to exercise some sort of control.

Some work has been carried out on horizontally growing stem systems such as rhizomes and stolons. When the direction of growth is at right angles to the direction of the stimulus, the response is described as *diatropic*, so that the response by growing horizontally at right angles to the gravitational stimulus is known as *diageotropism*. Potato provides a good example of a plagiotropic stem (stolon) which grows erect if buds are removed from the main shoots above ground. Control may be modified by external environmental factors as for example in the case of that troublesome weed *Aegopodium podagraria* (ground elder). Here the normally diageotropic rhizome becomes positively plagiogeotropic (i.e. it starts to grow downwards at an angle to the horizontal) if it becomes exposed to red light. If it is surrounded by air enriched with 5% of CO_2 it will turn upwards at an angle towards the surface. Thus there is a process of continuing adjustment of the growth alignment of this rhizome, so that in a normal loam soil it will maintain itself at a depth of about 10 cm below the soil surface.

PART FOUR

Reproduction in the working plant

14   The flowering process

14 · 1 *Growth and development at the flowering apex*

Reproduction in the flowering plant involves a highly co-ordinated sequence of events, starting with the induction of the flowering state in the meristematic stem apex, and leading right onwards to the dispersal of seeds from the parent plant. Description of the induction process will be deferred for the moment because it involves a number of related phenomena which are best dealt with at a later stage.

The sequence of the various floral primordia referred to in 12 · 2 may differ somewhat from flower to flower. The logical sequence of sepals, petals, stamens and ovary is seen in many plants but is by no means invariable; thus petal rudiments may sometimes be the last to differentiate and to start growing. Various degrees of fusion are shown between and amongst these different whorls. Thus to take examples from one family, the Caryophyllaceae, in all members of one tribe, the Alsinoideae (including stitchwort and chickweed), the sepals remain as distinct and leaf-like outgrowths (Fig. 14 · 1a and b). In the related tribe of the Silenoideae (including the pinks and campions), the margins of the calyx or sepal rudiments cohere and fuse as they grow up, and are only separately distinguishable at their apices (Fig. 14 · 1c). In the related primrose family (Primulaceae), the corolla whorl also consists of fused members forming a corolla tube. The developing stamens are adherent to the inside of this tube, and they are carried up with the corolla tube to a greater or lesser extent. The flowers are described as thrum-eyed if the stamens are at a level above the stigma, and pin-eyed if they arise below the level of the stigma (Fig. 14 · c). Growth here (as frequently also seen in the ovary) takes the form of meristematic activity on the rim of what is essentially an extending tube, and this type of growth is sometimes referred to as *toroidal*. In a *hypogynous* flower, as the name suggests, all other flowering parts (stamens, petals, sepals, in that descending order) are inserted below the *superior* ovary (Fig. 14 · 2a). If a *torus* or receptacle develops

177

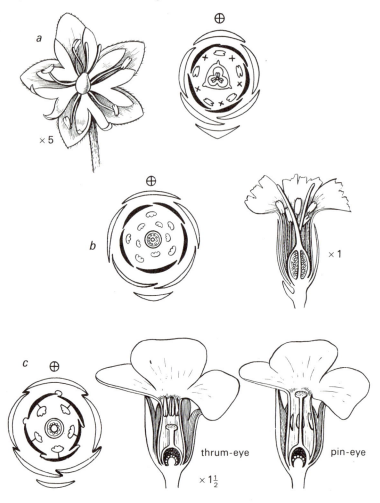

Fig 14·1 Aspects of flower structure 1: fusion of sepals and petals.
a. Stellaria media (chickweed) Caryophyllaceae
b. Dianthus sp. (pink) Caryophyllaceae.
c. Primula sp. (primrose) Primulaceae.

The floral diagrams enable one to show with formal brevity the interrelation of the various floral whorls and also some of their characteristics. Thus one can read from the diagram of *Primula sp.* that sepal and petal whorls are each fused to form tubes, that the anthers dehisce inwards and are opposite and fused to the petals, and that the placentation of the ovary is free-central. The combination of a floral diagram with its complementary median longitudinal section makes available a great deal of information about the way in which a flower works.

178

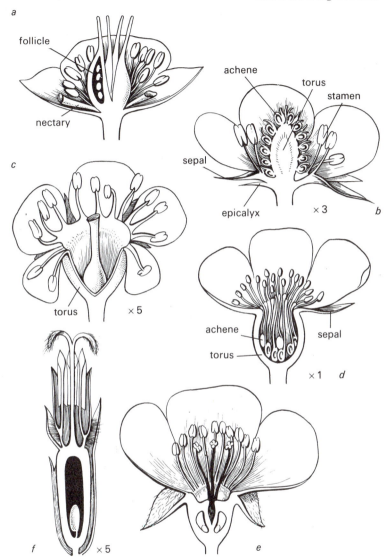

Fig 14·2 Aspects of flower structure 2: position of the ovary.

a. Hypogyny in *Helleborus sp.*, Ranunculaceae.

b. The first stages in perigyny in *Fragaria sp.* (strawberry) Rosaceae.

c. Perigyny in *Prunus sp.* (cherry) Rosaceae.

d. Perigyny in *Rosa sp.* (rose) Rosaceae.

e. Epigyny in *Pyrus malus* (apple) Rosaceae.

f. Epigyny in *Helianthus annuus* (sunflower) Compositae; a single floret taken from

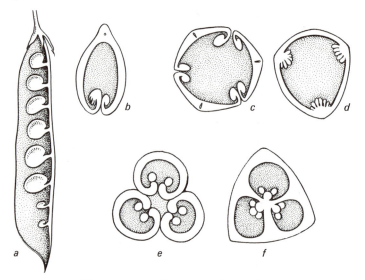

Fig 14·3 Aspects of flower structure 3: the carpel theory of ovary structure.
a. A folded leaf with fertile margins, fusing to give a legume as in *Pisum sativum* (pea) or to give follicles as in the apocarpous ovary in *Helleborus sp.* (cf. Fig 14·2a) or in *Delphinium sp.* as shown in *b*.
c. Three carpels fuse by their margins and grow to form a syncarpous ovary with parietal placentation, as in *Viola sp.* (*d*).
e. A syncarpous ovary with axile placentation, as based on the carpel theory (see *f Lilium sp.*)

by ring-growth, these parts may arise at or about the same level as the ovary. This type of flower is known as *perigynous*, and reaches its extreme in the rose, where the torus does not quite fuse over the ovary (see Fig. 14 · 2b, c and d). Where fusion takes place, and calyx, petals and stamens are borne on top of the ovary, the flower is described as *epigynous* (14 · 2e and f). It may well be imagined what tightly-integrated sequences of tissue development are involved in these processes.

Ovary growth may be even more complex. It is fairly simple to picture an *apocarpous* ovary, in which separate carpels (fertile leaves) sit at the apex of the floral axis (Fig. 14 · 2a and 3a). It can be imagined that carpels arise by the folding together of fertile leaves, whose fertile margins fuse together as in a pea pod (14 · 3b and c). It is also fairly simple to understand the possibility of say, three leaf-like carpels fusing by their margins to give a compound *syn-*

carpous ovary, in which the fused margins are fertile and bear ovules (Fig. 14 · 3d). The manner and position in which the ovules arise is known as the *placentation* of the ovary and the arrangement described in Fig. 14 · 3d would be called as *parietal* placentation. Figure 14 · 3e shows one way in which carpel theory can be held to account for the axile placentation that is seen in a lily ovary. However, it is much more difficult to account for the growth of an *inferior* ovary, where all the parts of the epigynous flower surmount the ovary beneath them. Nevertheless it is not proposed to do more here than to call attention to the kinds of growth adjustment that have to be made, and to observe that these really are of secondary importance to the two major events of flowering, namely (*a*) spore formation, following upon meiosis or reduction division and (*b*) the formation of male and female gametes, fusion of which gives rise to the zygote which develops to an embryo, and hence to a new organism. The stamen gives rise to microspores or pollen grains, from which are derived male gametes. The ovule similarly develops an embryosac, corresponding to a megaspore, and within this sac is developed the ovum or female gamete.

14 · 2 *Anthers and pollen*

At the floral apex, each of the rudiments that gives rise to a stamen starts to grow in length by apical growth, and the papilla that first develops is rod-shaped. It may show the development of a *filament* or stalk, and this is surmounted by the *anther* which is broadened by marginal growth. Investigation shows that in each of the four corners of the anther a group of *archesporial* cells becomes delimited (see Fig. 14 · 4). Each of these zones gives rise to potentially sporogenous tissue, and it is not long before pollen (or microspore) mother cells arise, the nuclei of which undergo meiotic division resulting in four haploid daughter nuclei; around each of the daughter nuclei is organised a microspore or pollen grain. The developing pollen grains are nourished in part by the breakdown of the nutritive and secretory layer known as the *tapetum*, and they come to lie freely in a pollen sac that is filled at first with a nutritive mushy fluid. As they mature the grains develop a layered wall, of which the outer composite layer (*exine*) is variously sculptured and cutinised, and may show more or less prominent pores.

During this period cellulose thickenings have been laid down in the form of radial bars in the *fibrous layer* of the anther (see Fig.

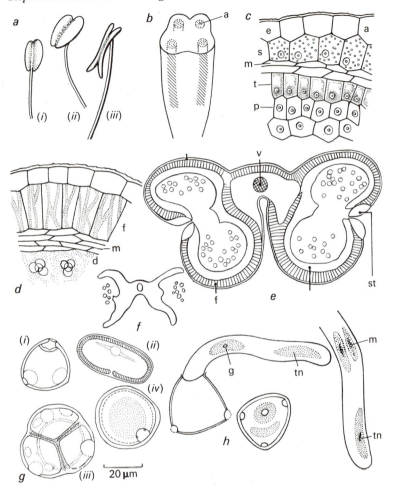

Fig 14·4 The anther and pollen production.
a. Stamen types; (*i*) basifixed, in which the filament is continued upwards into the connective which joins the two halves of the anther (e.g. *Ranunculus acris*, meadow buttercup); (*ii*) dorsifixed, in which the filament joins the connective halfway along the anther (e.g. *Cheiranthus cheiri*, wallflower); (*iii*) versatile, in which the junction of the filament and the connective is flexible, and the anthers move in the wind (e.g. *Lolium perenne*, rye grass).
b. A view of half of an anther primordium at the archesporial (a) stage.
c. T.S. of part of an anther lobe, showing the epidermis (e), starch layer (s), middle layer (m) and tapetum (t) just before meiosis in the pollen mother cells (pm).
d. After the differentiation of pollen, the cell wall structure develops a fibrous layer (f) and the tapetum disorganises (d). (Continued on opposite page)

14 · 4d) and these cause the strains due to drying out to be directed around the periphery of the anther, and to operate on lines of weakness in the wall corresponding to the slits which subsequently develop. Withdrawal of water from the anther may be assisted by the action of nearby nectaries, but its effect is always to accentuate radial stress in the anther walls in such a way that there is breakage at the predetermined lines of weakness.

By now the pollen grains are lying fully formed within the pollen sac and the bulk of the nutrient material has been absorbed by the grains as they mature. The mass of pollen gradually dries out, and by the time that the anther bursts open, the pollen is exposed and ready for dispersal.

In *introrse* dehiscence the pollen is shed inwards towards the stigma(s), making self pollination more likely, whereas in *extrorse* dehiscence the pollen is shed outwards; in some flowers the stamens dehisce laterally.

Usually during germination of the pollen grain, the *intine* (or innermost cellulose layer of the composite wall) grows out through one of the pores. Sometimes more than one pollen tube may be formed, though generally one only grows down through the stigma. Growth of the pollen tube is by tip extension, just as in a root hair.

Characteristically, the pollen grains at the time of shedding may have only two nuclei, but sometimes a second nuclear division may take place before the grain is shed. Of the first two nuclei, one ultimately degenerates; this is known as the *vegetative* or *tube* nucleus (though there is no solid evidence to confirm that it really controls the germination and growth of the pollen tube). The other nucleus is known as the *generative* nucleus, and this divides to give two *male nuclei*, one of which eventually fuses with the ovum within the embryo sac.

Wind-borne pollen is generally small, light and smooth walled. Insect-transmitted pollen is often relatively larger and heavier, and can thus contain more reserves for the subsequent germination

e. T.S. of a mature anther just before dehiscence, showing the vascular supply (v) to the connective (c), pollen grains (p) in the pollen sac, and the stomium (st), or line of dehiscence.
f. A dehisced anther.
g. Pollen from (*i*) *Corylus avellana* (hazel), (*ii*) *Heracleum sphondylium* (hogweed), (*iii*) *Calluna vulgaris* (ling) pollen tetrad, (*iv*) *Phleum pratense* (timothy grass).
h. Stages in pollen germination, showing the generative nucleus (g), tube nucleus (t) and male nuclei (m).

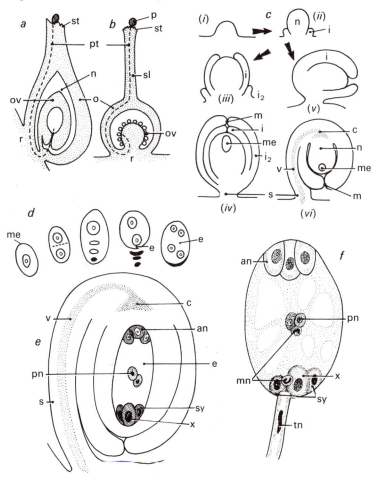

Fig 14·5 The ovule and fertilisation.

a. An achene with a single ovule.

b. A multiovulate syncarpous ovary, with axile placentation, showing the stigma (st), style (sl), a germinating pollen grain (gp) with its pollen tube (pt) the receptacle (r), an ovule (ov) inside the nucellus (n) and ovary wall (o).

c. Stages in the development of an ovule.

(*i*) A papilla arises on the placenta and becomes the nucellus (n) (*ii*) The first integument grows as a collar round the base (i). (*iii*) An orthotropous ovule with two integuments, outer (i_2) and inner (i). (*iv*) The same at megaspore mother cell stage (me). (*v*) An intermediate stage in the development of an anatropous ovule. (*vi*) The same at the megaspore mother-cell stage, showing the vascular supply (v), the chalaza (c), the micropyle (m) and the stalk (s). (Continued on opposite page) and early growth stages of the pollen tube. In its later stages of growth, the pollen tube has to be nourished by the stylar tissues of the plant receiving pollen, as it grows through them on its way towards the ovule. Such insect-transmitted pollen is often more heavily sculptured and ornamented (even spinous) and it clings better to the frequently papillose stigmatic surface, which is also sticky with secreted sugary material.

14 · 3 *Ovules and embryo sac*

Reference may be made throughout this Section to Fig. 14 · 5.

As already seen, the ovary, bearing its ovules on placental areas, may consist of a single carpel as in the Leguminosae, a number of separate carpels forming an *apocarpous* ovary as in the Ranunculaceae, or from two to many fused carpels forming a syncarpous ovary, as in gooseberry, tomato, tulip etc. The carpel or the ovary walls protect the developing ovules within, and after fertilisation of the ovules may themselves develop further to form the fruit. The tip of the ovary is developed more or less to form a sticky or a hairy receiving surface, the *stigma*, which may be raised on a stalk-like or columnar *style*. Pollen grains landing on the receptive stigmatic surface may germinate if conditions are favourable, and the pollen tube grows downwards through stylar tissue towards the ovules within the ovary. Ovules start to develop as small papillae which arise from placental tissue. The growing papilla becomes invested with one or two *integuments*, which develop by toroidal growth from the base of the shortly stalked papilla. There is now a *nucellus* surrounded by one or two protective envelopes, the integuments, which grow on upwards until they all but meet over the top of the nucellus, leaving a small channel known as the *micropyle*. Within the nucellus, one selected cell (the megaspore or embryo-sac mother cell) undergoes meiosis and generally produces a chain of four

d. A diagram of stages in the development of the embryo sac (e) from the megaspore mother cell. Only one haploid cell survives from the linear tetrad of cells resulting from meiosis. The combination of synergids (sy) (supporting cells) and ovum (x) is known as the egg apparatus: centrally, the polar nuclei are seen (pn) and at the opposite end are the antipodal cells (an) which at first have no cell wall.

e. An anatropous ovule ready for fertilisation.

f. The embryo sac at the time of fertilisation; one male nucleus fuses with the egg, and the second with the two polar nuclei to give a triploid product, the precursor of the endosperm.

185

haploid megaspores of which all but the innermost degenerate. The persistent megaspore enlarges to become the embryo-sac and this is at first uninucleate. However, the single nucleus shortly divides by three successive mitotic divisions to give eight haploid nuclei. Figure 14·5 shows the characteristic arrangement of these nuclei when the ovule is ripe for fertilisation. The three nuclei at the end of the embryo-sac remote from the micropyle (the so-called chalazal region) develop thin walls and are known as the antipodal cells. Of the three nuclei which migrate to the micropylar end of the embryo-sac, one acts as the ovum and the remaining two act as *supporting cells* or *synergidae.* This leaves one nuclear member from each quartet, and this pair of nuclei (one of which might have been an egg and one of which might have been an antipodal cell) associate closely together in the centre of the sac, and may even fuse before fertilisation. Together with one of the male cells which is shed into the embryo-sac from the pollen tube, they eventually fuse to form the single triploid *primary endosperm nucleus.*

The arrangement described is characteristic of about two thirds of the embryo-sacs studied, and the numerous variations on this theme cannot be our concern here. Similarly our consideration of the growth and orientation of the ovule must be restricted to the rather more common *anatropous* condition in which (as shown in Fig. 14·5) by curvature of the ovular stalk or *funicle,* the ovule finishes up by having its micropyle oriented so as to face backwards towards the placenta from which it has arisen. This has obvious advantages for a pollen tube that has grown down through style and ovary wall to the placental tissue, and has then only a small distance to bridge before entering the micropyle and penetrating through to the embryo-sac.

14·4 *Pollination*

This is the process by which pollen is transferred from the anther to the receptive (stigmatic) surface of the ovary. The mechanism of transfer may involve wind or water, but, more characteristically, modern angiosperms have adapted themselves to a form of co-operation with insects, by whose agency pollen is transferred from flower to flower. Many flowers are open to all-comers, but others restrict insect visitation to long-tongued insects; in the Arum lilies the visitors are temporarily imprisoned until their imposed mission is fulfilled. The adaptations involved in entomophilous (insect-pollinated) flowers include attraction by colour and scent, and the provision of rewards in the form of nectar or pollen.

A good deal is known about the perception of colour and patterns by insects, especially by bees. By associating food with different coloured backgrounds it is found that bees are unable to distinguish between reds and blacks and greys; they are red-blind. However, red flowers frequently reflect light in the ultra-violet and may be distinguished in this way. Other insects are more sensitive to parts of the red end of the spectrum. The flower colours involved are mostly gene-controlled; they may be sap-soluble pigments in the vacuole (such as the largely blue and red anthocyanin pigments) or they may be yellow to red chromoplastid pigments akin to carotene. Combinations of these two types of pigment give a wide range of colours. Normally colour production is restricted to the petal whorls, but there are many plants in which the calyx bears the same colour (e.g. lilies); sometimes the whole inflorescence is coloured, as in the culinary herb *Salvia officinalis* (sage) and sometimes the stamens and parts of the ovary are highly coloured. Notice also how with other families (Compositae and Umbelliferae) conspicuousness is achieved by the massing of many small flowers in one, often quite complex, inflorescence.

All too little is known about the fragrance of flowers. The production of scent can be a periodic phenomenon, as in the case of jasmine, tobacco and evening primrose, which seem to be at their most overpowering at dusk. In some cases this is undoubtedly linked with visitation by certain nocturnal moths. Scent production certainly reaches a maximum just after *anthesis* (i.e. as the flower is opening), and this represents a subtle form of timing in the production of highly specific chemical molecules within a limited and accurately located period of time.

14 · 4 : 1 Aspects of pollen physiology

Whilst in some plants the pollen grains will germinate in water only, sugars and organic acids have been identified in the stigmatic fluid as substances which stimulate pollen tube growth. In many other plants not only must sugar be present for germination but it must also be present at fairly specific concentrations. It is wise when trying to germinate pollen of such plants as tulip or daffodil to use a series of sugar concentrations between 0 and, say, 15%. It has been found, too, that traces of boron are necessary for pollen germination, though it is not really known how boron works. It is possible that it intervenes in the synthesis of pectic substances used for building the tube wall. It is also reported that germination is more pronounced when pollen grains are clumped together, and it is thought that this may be related to the overall concentration of calcium that is available as it diffuses out from the grains in the early stages.

The rate of growth of a pollen tube varies considerably from species to species. In *Crocus* the pollen tube may travel a distance of between 6 and 10 cm down the style in anything from one to three days. In birch, from pollination to fertilisation takes one month, and in some species of oak it may be as long as a year or more.

The growth rate of a pollen tube may be retarded in various ways during its journey. The stigma may be provided with a thin membrane of cutin, and unless the pollen tube can provide cutin-splitting enzymes, the delay in growth may cause a time barrier to fertilisation by that particular grain. Other barriers may include inhibitory substances secreted by the ovule—in this case providing metabolic barriers to growth. There is a good deal of evidence to suggest that a protein diffusing from the pollen tube may act as an antigen-like substance combining with and inactivating an antibody manufactured in the style. If the latter accumulates to a sufficiently high level (in an incompatible pollination) the respiration rate goes up, and the stylar tissues seem no longer able to provide the supply of amino-acids normally available for the growth of the developing tube.

Pollination has been clearly shown to be involved in the setting of fruit. If unpollinated tomato plants are treated with IAA, the fruits set and develop, even though no fertile seeds are formed within the ovary. In the absence of pollination or auxin treatment, the unfertilised fruits will wither and drop off. Thus pollination appears to stimulate the beginning of ovule expansion, which is of course under the influence of fertilisation and the development of the ovum. Pollination also inhibits the development of the abscission layer at the base of the ovary, thus allowing the stem to thicken and support the developing ovary. The formation of fruit in the absence of fertilisation is known as parthenocarpy.

One final point is worth a brief mention. Using the techniques of tissue culture, it has recently been demonstrated that growing pollen grains can be induced, like many other active cells, to give rise to embryoids or to undifferentiated callus tissues. Both of these can be made to yield plantlets (as with Steward's carrot plants) but they carry the haploid number of chromosomes. Such plants are of course infertile, but under chemical treatment with the alkaloid colchicine at an early stage in the process the diploid state may be induced, and the diploid plants that result from this treatment are fertile. Each has a genome that is based upon a selection of the parental characters which underwent reassortment at meiosis and appeared in each individual pollen grain from which plants are derived. This is a technique which could be of some value in plant breeding programmes, since the new plants are homozygous for all of the characters that they carry. Obviously a lot of screening remains to be done to find out which of these plants carries desirable characters.

A further advantage lies in the shortening of the time scale of the breeding programme. In tobacco plants, for example, it is thought possible to produce homozygous seeds in something under six months. This could be very valuable where the plant concerned has a short generation time, though progress is very much slower where, as with the oil palms, the generation time is so prolonged.

14 · 5 *Breeding behaviour*

By *outbreeding* is meant the situation in which pollen from a plant of a given genotype is transferred to the stigma of a plant with a differing genotype. In inbreeding a plant is pollinated by its own pollen (*autogamy*).

Male nucleus and ovum each contribute a haploid set of chromosomes to the diploid zygote. No new factor can be introduced into the diploid genotype if each contributory haploid set of genes is identical, though of course fresh groupings of existing characters can arise through random separation of chromosomes, through crossing over during meiosis and through random combination during fertilisation. (See Ashton's *Genes, Chromosomes and Evolution* in this series.) Only through outbreeding can new genes (including mutations) be introduced into the genotype, and selection then operates in such a way as to preserve plants exhibiting those features which best help the population to survive. Outbreeding makes it possible for there to be a full sampling and exchange within the genotype (or pool of genes) that exists in a given population.

The process of evolution does not stand still, and populations of plants that cannot adapt themselves to environmental change must make place for those that can. The full complement of genes that determine the range of form and function of the plants in a population, must itself be capable of modification, if adaptation is to be such as to ensure a good chance of survival.

We therefore have to be on the lookout for any factor in flowering which makes it more likely that the ovules of a given plant will be fertilised by gametes from the pollen of a different plant. The chief of these factors may involve the physical separation in space or in time of ripe ovules and viable pollen in one and the same plant (dichogamy). For example, pollen may be shed well before the stigmas on the same plant are ripe; such plants are said to be pro-tandrous, and *protandry* is shown by a number of families, e.g. the Compositae, Caryophyllaceae, Geraniaceae etc. *Protogyny*, in which the stigmas ripen before the pollen is shed, may be seen in the plantains (*Plantago spp.*) and a number of common grasses. It will be recognised that separation of the sexes of this kind arises through control of the developmental sequences leading to stamen and ovule formation.

Amongst the willows (*Salix spp.*), whose flowers produce nectar and are insect pollinated, one finds a separation of the male and female catkins on distinct plants. Such plants are described as *dioecious*. Most of the wind pollinated trees (such as poplar, oak, beech, birch etc.) are *monoecious*; groups of male or female flowers are separated from one another though both are borne on the same plant, and these plants are generally protandrous.

Thus in outbreeding we must reckon also with the factor of sex-determination. In a normal hermaphrodite flower, 'maleness' is expressed in stamen formation leading to pollen production and the provision of male nuclei for fertilisation. 'Femaleness' is expressed in the production of ovules within an ovary. In cucurbits such as cucumber or marrow, a long trailing stem initiates a sequence of axillary flowers in which the first and furthest from the apex are normally all male (i.e. the flowers produce stamens but the ovaries do not develop). Further up the stem femaleness becomes more pronounced and stamen formation is suppressed. Intergradations are to be found between these extremes.

The prevailing level of IAA within the plant seems to be of some importance in determining sex expression. Under conditions in which auxin levels are depressed, a plant like marrow responds by producing a larger proportion of male flowers. These include such conditions as a shorter daylength (see p. 199), higher night temperatures, the presence of carbon monoxide and the application of gibberellin. If auxin is applied in a paste to the apex of a marrow plant more female flowers are formed, whilst the application of gibberellin appears to promote maleness. Thus we can glimpse a little of the means of control by means of which a separation in time of male and female plants can be achieved, and this makes outbreeding all the more likely.

A final comment should be made on the way in which plants of different families have adapted themselves, during the course of evolution, to co-operation with the insects which have been evolving side by side with them. Insect-pollinated flowers may be divided into a number of classes, including:

1. those flowers which offer only pollen to the visiting insect (e.g. the poppies)
2. those with freely exposed nectar (e.g. buttercup)
3. those with partly-concealed nectar (e.g. buckwheat—*Fagopyrum*)
4. flowers with fully-concealed nectar (e.g. members of the Compositae, which also show a massing of small flowers to form a conspicuous attractive inflorescence or *capitulum*).
5. Lepidopteran flowers, whose long corolla tubes conceal nectar which can only be reached by long-tongued butterflies or moths.
6. bee flowers, often very irregular with well-concealed nectar, and often exhibiting a floral mechanism which requires the presence of a long-tongued insect with the size and weight of a bee (e.g. many Leguminosae, Labiatae etc.).

There is a tendency, whilst going through the exercise of identifying a flower with the use of a key, to overlook the means by which the flower copes with the problem of pollination and outbreeding. It is always worth while to stand aside a little during such an exercise, and to consider the flower in its deeper biological context as an adaptive mechanism for the achievement of fertilisation. It is useful to speculate as to how far its adaptation has been ' successful ', and to try to think out some of the criteria by which ' success ' might possibly be assessed.

14 · 6 *Fertilisation and its consequences*

Once pollination has taken place and pollen tubes are growing chemotropically towards the ovules, the way is open for the process of fertilisation, i.e. for the fusion of gametes within the embryo-sac. The pollen tube grows down the stylar tissue and ovary wall, emerging to bridge the short gap to the micropyle of the ovule. Presumably, as with fungal hyphae, the tip of a pollen tube secretes enzymes which promote the hydrolysis of middle lamella substance, and this enables the tube to separate (and pass between) cells of the ovary tissue. Probably other hydrolases act to provide the simple nutrients that the pollen tube requires for growth en route, and we have already seen how action and reaction between tube and style may make amino-acids available. We may also presume that the stimulus which directs tube growth, ceases after fertilisation, and that the pollen tubes following grow on towards other as yet unfertilised ovules.

The pollen tube enters the embryo-sac and ruptures; the two male gametes, each consisting of nucleus plus a little cytoplasm, move into the sac where one fuses with the ovum. The synergid cells which support the ovum may initially have a nutritive function, but they soon disappear after fertilisation. The second male nucleus fuses with the two polar nuclei at the centre of the embryo-sac, though these may sometimes already have undergone fusion. The final product, which is the fusion product of three nuclei is therefore triploid and is known as the *primary endosperm nucleus.* In one group of plants this develops by mitosis to a free-nucleate condition; there are few cell walls and the endosperm is fairly short lived, disappearing gradually as the embryo develops in size and takes over the function of food storage. In other plants (e.g. buckwheat) cell walls are laid down and the endosperm becomes a prominent food-storing tissue surrounding the embryo. Amongst the mono-cotyledons, the presence in the cereal grasses of a starchy endosperm is of great economic importance. (It is customary and useful to distinguish in this way between endospermic and non-endospermic seeds. In a very few instances food is stored in nucellar tissue, but in most non-endospermic seeds the reserves are stored in the cotyledons.)

14 · 6 : 1 Development of the embryo

For the last century, *Capsella*, the shepherd's purse, has been used to provide illustrative material of stages in the development of the dicotyledonous embryo, and the stages described here all refer to *Capsella* (Fig. 14 · 6). The fertilised ovum secretes a wall, elongates

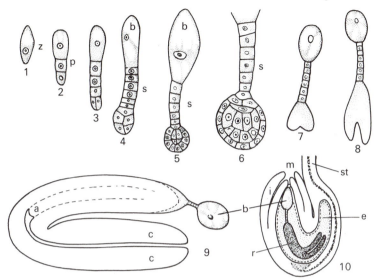

Fig 14·6 Stages in the development of the embryo in *Capsella bursa-pastoris* (shepherd's purse). 1–6 after Souèges; 7–10 after Schaffner. 4, octant stage; 5, the periclinal walls delimit the dermatogen; 7, the cordate or heart-shaped stage; 8, the embryo begins to lengthen; 9, the embryo is fully formed; and 10, the embryo within the ovule.

Abbreviations; z, zygote; p, proembryo; b, basal cell; s, suspensor; a, stem apex; c, cotyledon; r, radicle; e, embryo sac; i, integument; m, micropyle and st, stalk.

somewhat, and then divides by a wall at right angles to its long axis giving rise to a *basal* cell at the micropylar end of the embryo-sac and a *terminal* cell. The basal cell by transverse divisions becomes an elongated filament (the suspensor), with a swollen terminal cell which seems to anchor the elongating suspensor filament, thus thrusting the distal terminal cell of the pro-embryo deep down into the nutrient ' mush ' of the embryo-sac.

The terminal cell divides first by anticlinal walls (at right angles to the surface) to give eight cells known as the ' octant ' stage (Fig. 14·6, 4). Walls are then laid down periclinally, and delimit a central mass of cells from an outer series of surface cells which constitutes the dermatogen or embryonic epidermis. Subsequent divisions result in a somewhat undifferentiated-looking spherical mass of cells (sometimes called the ' mulberry ' stage). However, by subsequent differential growth two cotyledons are defined, together

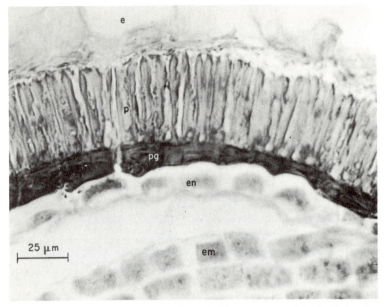

Fig 14·7 A section through the testa of a seed of *Sinapis arvensis* (charlock) showing the mucilaginous epidermis (e), the lignified palisade layer (p), the pigmented layer (pg), the remains of the endosperm (en) and part of the embryo (em).

with a hypocotyl and a root primordium which is immediately next to the suspensor. The developing embryo grows to occupy the embryo-sac cavity.

Meanwhile the ovule wall is undergoing growth changes which enable it to accommodate itself to the developing embryo, and to develop the protective testa or seed coat (Fig. 14 · 7).

The developing embryo (and maybe some of the tissues around it) appears to be a potent source of substances which stimulate growth of the ovary wall (or *pericarp*), though, as we have seen, the stimulus of pollination may also contribute to this phase of development.

14 · 7 Development of fruits and seeds

The older literature is well provided with morphological descriptions of types of fruits and seeds, and their nomenclature. Here we will refer only briefly to this aspect whilst concentrating more on the biological principles involved. The sequence of pollination, fertilisation and the growth of seeds within a protective ovarial wall draws attention to the fact that naked ovules are practically unknown amongst the *Angiosperms* or flowering plants. For an exception look at the flowers of mignonette (*Reseda* spp.), but ovules are nearly always enclosed or protected within ovary walls. In the *Gymnosperms* the ovules lie on an unfolded open ovuliferous scale; in practice the closely imbricated cone affords almost as good protection as an ovary wall. Food materials must reach the flowering plant ovule via the vascular system of the ovary walls. As the seeds develop within the ovary the latter may develop in a wide range of ways. It may become excessively fleshy, and may show (as in a plum) an outside skin (*exocarp*), a pulpy *mesocarp* and a hard *endocarp* (the stone of the plum). In the olive the fleshy mesocarp stores oil, which is extremely important in the food and other industries. On the other hand the ovary may become inextricably joined up with receptacle tissue, and the resultant so-called false fruit may best be exemplified by the apple or the pear. Such fleshy fruits are mostly animal dispersed and the seeds, if not dropped by the animal almost at once with the remains of the fruit, may pass through the digestive tract and be dropped later (*endozoic* dispersal). In either instance the seeds germinate at some distance from the parent plant. Most fleshy fruits consist of tissues which, physiologically speaking, are past the point of no return and can never regenerate if cultured. These tissues are senescent and as they decline towards their death, they reach a state that is termed the *climacteric*. At this point their respiration rate is generally high, and their production of volatile by-products confers flavour and attractiveness in them. We now know that ripe fruits give off traces of ethylene (of the order of one part per million), and that this small quantity of ethylene can speed up the ripening of other fruits in the vicinity. The fruit trade is dependent on storage techniques, such as the use of lowered temperatures and raised CO_2 concentrations, for delaying the achieval of the climacteric in order that fruits that are released on to the retail market may be at their best in flavour and texture.

Another group of fruits have an ovary wall that progressively dries out and hardens without splitting (these are referred to as *indehiscent* fruits). Sometimes the hard seed-containing ovary wall (as in the hazel nut) has to lie in the ground until bacterial action softens the wall sufficiently to allow the entry of water and the swelling of the seed within. Sometimes the ovary wall is provided with hooks, e.g. *Circaea lutetiana* (Enchanter's Nightshade), which provide a means of *epizoic* dispersal on the outside of an animal. *Dehiscent* fruits open in a variety of ways, often drying out so that the ovary wall is under tension, and splitting with such violence as to release and scatter the seeds away from the parent plant.

Thus the post-fertilisation development of the ovary, as distinct from the fertilised ovules that it contains, is concerned firstly with the protection of the ovules during growth, and later with the dispersal of the seeds into which they grow. Sometimes the ovary wall is also concerned with delay in the germination of the seed (see p. 212), largely because of the presence of substances which inhibit growth.

Seeds, as has already been stated, are with few exceptions either endospermic or non-endospermic. Before dormancy sets in food reserves are laid down either in the tissues surrounding the embryo, or within the embryo itself. In endospermic seeds food must be mobilised and transferred across to the embryo for its subsequent growth. For example, the cereal grasses have developed a special organ, the *scutellum*, which is part of the embryo and lies between it and the starch endosperm. It is now well established that gibberellin from the imbibed embryo stimulates the production of starch- and protein-splitting enzymes; by their action, sugars and amino-acids are made available for transport to the developing embryonic issues via the scutellum. (See Fig. 13 · 1a).

It must be obvious that the amount of food stored in a seed may be crucial to its survival and subsequent growth, often in the face of competing neighbours. Some of the smallest and most immature seeds are found in the orchids, where the seed as it is shed consists of little more than a rod of undifferentiated tissue, surrounded by a thin papery testa. Whilst this makes wind dispersal of the seed very effective, because it is so small and light, there are little or no reserves, and the seeds are dependent for their nutrition on their symbiotic relationship with a fungus. The fungus makes sugars available from the substrate to the diminutive embryo, which may take months to develop to the point of being capable of growth and independent survival. Nowadays growers may surface-sterilise such seeds and grow them on in sterile culture on the surface of a nutrient agar containing the requisite sugars, and thus speed up the whole process. An even more useful application of the techniques of tissue culture consists in developing orchid cell cultures, within which, by appropriate treatments, embryoids and later on mature plants may be grown. This makes it possible to produce non-variant clones of any particular plant that has caught the fancy of the enthusiasts.

The chances of survival and success do clearly depend on food stores in the seed, and for the most part these include starch, proteins and oils. The materials stored must be osmotically inactive; the substances that contain the least amount of oxygen in their molecules will provide the greatest amount of energy during germination. If we compare the energy available per gram of hexose (derived from starch) with that provided by fatty materials it is clear that fats are more efficient storers of energy. Hexose can yield energy of the order of $15.5\,Jg^{-1}$, whilst oils may yield as much as $39.9\,Jg^{-1}$. Proteins yield intermediate amounts of the order of $23.9\,Jg^{-1}$. Quite a large amount of this energy is of course dissipated as heat during respiration. It should be recognised that a high percentage of the world's staple food products are based essentially on seed storage. The cereals and legumes are outstanding in this respect, but oil seeds such as rape, coconut, cotton etc. must not be left out of account for they are of value not only as food, but are also extensively used in the soap and paint industries.

15 Some physiological aspects of flowering

15 · 1 *Nutrition, temperature and light*

We have known now for more than half a century that plants pass from the vegetative to the flowering condition for reasons which are sometimes connected with light, sometimes with temperature, and sometimes with nutrition and the water regime.

Some of the early work on this problem suggested a causal relationship between flowering in a plant and its nutritional status. In particular it was thought that flowering might depend on the balance in the plant between the levels of carbohydrates versus nitrogen compounds. It is now recognised that the balance between vegetative and reproductive growth is a good deal more subtle than can be discerned by mere analysis of carbohydrate and nitrogen fractions.

Retardation of growth may bring a plant into flower at an earlier stage than usual. It is a fairly common observation that plants that are clinging on to life in a hostile situation may go over to flowering, having made a minimum of vegetative growth. Look, for example, at weeds such as *Senecio vulgaris* (groundsel) and *Capsella bursa-pastoris* (shepherd's purse) on any dry, paved walk, and compare their life span and earliness to flower with similar plants growing more normally in open soil nearby. The same sort of observation may often be made on the ephemeral plants of sand dunes.

It is similarly recognised that temperature may play a considerable part in flowering. This feature will be dealt with elsewhere as appropriate (see, for example, the paragraph on vernalisation (p. 207) and on the part played by temperature in the flowering of biennial plants). At this point we need only note that seasonal fluctuation in temperature may have a thermo-periodic effect, and, for example, that plants such as cabbage may be prevented from flowering if they are grown under conditions where the temperature is maintained at too high a level.

The American workers Garner and Allard demonstrated in the early 'twenties that the tobacco variety Maryland Mammoth would not flower at the normal June flowering period. However, if this variety were brought into the greenhouse and protected against cold, it could be induced to flower under the short days of winter. Plants which had remained vegetative during the long summer days could be caused to flower by cutting down their daily hours of exposure to light. The duration of exposure to light in any one day is known as the photoperiod.

This work led to the recognition that, whilst quite a large group of plants, including the sunflower, are seemingly indifferent to photoperiod, others may respond to daylengths of more than a critical minimum number of hours by flowering, e.g. the long-day plant *Hyoscyamus* (henbane); alternatively some may flower only when the hours of light do not exceed a certain critical maximum (e.g. short-day varieties of *Chrysanthemum*). Nowadays this early rather simple categorisation of flowering has been widely extended and it is recognised that *photoperiodism* is a phenomenon of widespread application and importance, influencing not only the induction of flower initials, but also the development of fruits and seeds, the vegetative development of tubers, the onset and break of dormancy in seeds and buds, and many other aspects of growth and reproduction.

15 · 2 *The phytochrome mechanism*

Borthwick and Hendricks and their colleagues in Maryland have demonstrated that the means by which plants perceive and react to changes in daylength depends in large measure upon an on-off switching mechanism involving the light-sensitive pigment phytochrome. Phytochrome is made up of a water-soluble protein molecule coupled with a pigment molecule that is most closely related to pigments (the *phycocyanins*) found in the blue-green algae. In the dark, phytochrome has a characteristic absorption spectrum showing absorption of red light with a peak at *ca.* 660 nm. On exposure to light, phytochrome changes its molecular structure (though it is not yet clear how), so that a peak of absorption now occurs in the far-red at 725 nm. This is a photochemical reaction of great

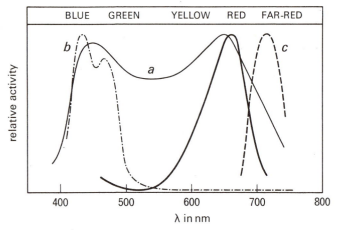

Fig 15·1 A graph showing the action spectra for *a*. photosynthesis
b. the growth response to light, and *c*. the phytochrome mechanism.

importance. It can be expressed conveniently as:

$$P_{660} \underset{\text{far-red light}}{\overset{\text{near-red light}}{\rightleftarrows}} P_{725}$$

[Attention should perhaps be drawn here to the fact that light influences plant activity in at least three main ways. Photosynthesis shows peaks of activity in the red and blue portion of the spectrum. Blue light (*ca.* 400–450 nm) is responsible for much of the growth response of plants to light, e.g. in phototropism, and is certainly implicated in the destruction of auxin. The third way involves the phytochrome near-red/far-red mechanism. Action spectra are indicated for these three mechanisms on the graph, Fig. 15 · 1.]

The phytochrome mechanism plays its part in a variety of happenings in the plant. It appears to be involved in the conversion of protochlorophyll to chlorophyll, and thus in the process of greening in the plant; it also has some control over the formation of the blue and red anthocyanin sap soluble pigments. In the process of germination it may be concerned either with the break or with the imposition of dormancy, according to the kind of seed being studied (see section 15·5). As the seedling emerges from underground, light from the near-red part of the spectrum operates, via phytochrome, to stimulate the unbending of the plumular hook (as for example in pea seedlings); it also stimulates the expansion of the

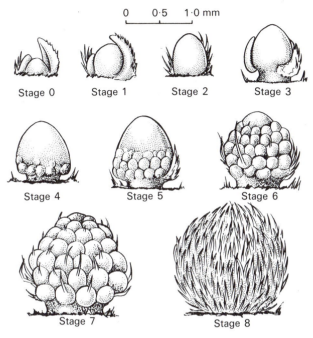

Fig 15·2 Some of the stages in floral initiation in *Xanthium sp.* (cocklebur) used to quantify floral induction.

leaves as they unfold on reaching the light. Not least, the phytochrome mechanism is fully implicated, as we shall see, in the photoperiodic response of the plant to daylength, or to put it in an alternative way, the response to the duration of darkness.

15 · 3 *The onset of flowering*

In testing out a light regime designed to accelerate or retard flowering, it is time-consuming to have to wait until the plant is actually showing flower. However, as we saw on p. 154 the growing apex reflects the first signs of change towards the flowering state, and it is often possible to shorten the investigation by careful examination after dissection of the growing point. In the case of the short-day plant *Xanthium* (the cocklebur), stages in floral initiation can be distinguished and have been used to provide a ' scoring ' method to indicate the progress of floral induction (see Fig. 15 · 2).

If short-day cocklebur has been grown under long days (e.g. a light period of 16 hours followed by 8 hours in darkness), then an exposure to *one* cycle only of short days (8L + 16D) is enough to promote flowering, and a few cycles of short days will set the plant flowering for two or three months or more, even if it continues to be kept under long-day conditions. Other plants require more cycles (i.e. more successive periods of exposure to light followed by dark) and the results may be cumulative in the sense that with increasing numbers of cycles the flowering response may be stronger and more firmly imprinted. However, in a plant such as Biloxi soybean, flower initiation stops as soon as the plant is returned into long-day conditions; the capacity of Biloxi soybean for ' remembering ' its previous experience is much more limited than that of *Xanthium*.

Just as short-day plants require not less than a given minimum period of darkness if they are to be induced to flower, so also long-day plants will not flower if the dark period exceeds a certain critical maximum.

Emphasis has been laid on the critical length of the dark period, because it has been shown in short-day plants that if the darkness is interrupted by exposure to light, the effects of the previous short-day exposure may be nullified. It rather looks as if a process is taking place in the dark which is counteracted by light. Reference to Fig. 15·3 will illustrate some of the main features of flowering in

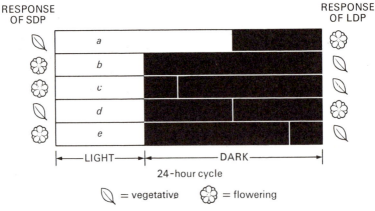

RESPONSE OF SDP

RESPONSE OF LDP

←—LIGHT—→←————DARK————→

24-hour cycle

🍃 = vegetative 🌸 = flowering

Fig 15·3*a* and *b*. Basic features of the responses of short- and long-day plants to day length.
c, *d*, and *e*. The results of interrupting the dark period with a flash, as discussed in the text.

relation to light and darkness. The basic differences between short-day and long-day plants are demonstrated in Fig. 15 · 3(a) and (b), though they can also be expressed in another way as shown in the graph (Fig. 15 · 4).

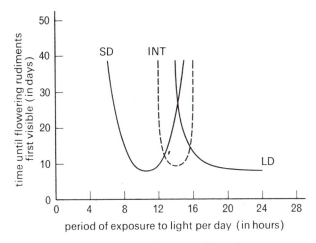

Fig 15·4 A diagram illustrating three types of flowering response to day length, in short-day (SD), intermediate (INT) and long-day (LD) plants.

If the short light flash is applied near the beginning of the dark period (c), it may reduce flowering only a little; if applied during the middle of the dark period (d), floral induction does not take place. If applied at the end of the dark period (e), once again flowering is little affected. It seems as though some unstable substance accumulates in the dark which may be destroyed by a light flash. At the beginning of the dark period not enough has accumulated for the destructive light flash to have made any difference. By the end of the dark period it seems to have been converted into another stable substance which can withstand the effects of the interrupting flash.

The light interruption may be of very short duration provided that the light is fairly intense. On the other hand with a commercial crop like chrysanthemums, flowering can be deferred almost indefinitely by leaving a very weak source of light shining on the plants during the night. It is of course essential that any plant should receive a sufficient quantity of light energy (duration ×

intensity) in order that it may carry out adequate photosynthesis for growth.

Figure 15·5a, b and c show the essential implication of the phytochrome mechanism in this process. An exposure to near-red light has the same effect as white light if applied in the middle of the dark period. Application of a series of flashes of light over a range of wave-lengths makes it possible to plot an action spectrum; a wavelength (λ) of 660 nm is the most effective (see Fig. 15·1). If, however, exposure to near-red is followed at once by an exposure to far-red

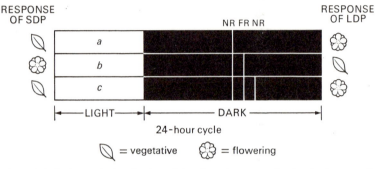

Fig 15·5 This shows the implication of the phytochrome mechanism in the case of plants whose dark period is interrupted by flashes of near-red or far-red light, as discussed in the text.

light ($\lambda = 725$ nm), the plants will flower; again, if near-red is followed by far-red light and this again by near-red, the short-day plants will stay vegetative. So, phytochrome in the far-red form, resulting from its exposure to near-red light, opposes the dark process, whereas a dark exposure or an exposure to far-red light promotes it. In true darkness, phytochrome reverts to the near-red form, just as if it had been exposed to far-red light.

This can be expressed diagrammatically:

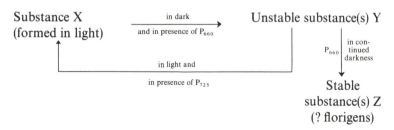

Thus in the few plants in which its presence is established and evidently linked to photoperiodism, phytochrome seems to be involved in a shunting mechanism that allows a dark reaction to go forward to completion only if the phytochrome stays in the near-red form. If exposure to light triggers the phytochrome back to the higher wave-length form (i.e. P_{725}) the process is stopped.

We have used symbols for the substances taking part in this process because we are ignorant of their true identity and indeed of the number of stages involved. However, we are sure that the process takes place chiefly in active green tissue i.e. mainly in the leaves. Young plants of cocklebur under long days are so sensitive that they can be photo-induced if all the leaves but one are darkened, and that one exposed to a single short-day cycle. In the short-day plant *Perilla* (one of the Labiatae or mint family), a Russian botanist has shown that the stimulus may be produced in as little as half a leaf, but that its effects are much reduced if it has to pass across a darkened area of leaf, that has previously been exposed to long days. It seems probable that the stable end-products of the process are one or more flower-induced hormones (some have called these *florigens*), and they are translocated from the leaf to the growing stem apex by normal phloem transport. It can be shown that the stimulus will pass across a graft union between two short-day plants, one of which has been exposed to short days and one of which has been kept on long days. The plant under long days is induced to flower even though it has been maintained under steady long-day conditions. Though the chemical nature of the transmitted substance is not known, it appears that it may pass between plants of the same variety or even between generically related plants of similar or dissimilar photoperiodic behaviour. For example, an induced plant of the succulent short-day plant *Kalanchoe* (SD) may pass ' florigen ' across a graft union to a long-day plant of *Sedum*. Other successful grafts which have been shown capable of transmitting the flowering stimulus were between *Xanthium* (cocklebur) and *Rudbeckia* (both composites) and between *Hyoscyamus* (henbane) and *Nicotiana* (tobacco). It is clearly essential that there should be contact between living cells for transmission to take place.

To summarise then: short-day plants need a minimum dark period in which a substance (or substances) formed in the leaves in the light, become converted to a relatively stable product; this is translocated out of the leaf to the vegetative apex, and there induces flowering. The dark process goes forward in the presence of phytochrome in the near-red form P_{660}, but the process can be blocked if illumination switches phytochrome to the far-red form P_{725}.

The behaviour of long-day plants is rather more difficult to explain, and frequently becomes confounded with other characteristics, in particular the need for a period of low temperature prior to flowering. Some of the most typical of the long-day plants are biennials, growing vegetatively in their first year, overwintering generally on a storage root or other organ, and then ' running to seed ' in the second season. The cold period of overwintering is an important part of the developmental process whereby the plant becomes ' ripe-to-flower '. Indeed, as already mentioned, plants such as cabbage or carrot can be kept in the vegetative state by maintaining them in a warm greenhouse and thus denying them the necessary cold period.

In contrast to short-day plants, long-day plants need less than a certain critical minimum length of exposure to darkness. If long-day plants are kept in the vegetative state by increasing their dark period beyond the critical minimum, a short interruption of the dark period by light (especially at 660 nm) will induce flowering. One hypothesis which could serve to explain this behaviour is that in these plants illumination by day results in a build up of labile florigen precursor; this stabilises quickly in darkness but goes on in the continued presence of P_{660} to produce a non-active (or non-mobile) derivative. Dozens of similar hypotheses could be made, but until we know the chemical nature of the hormone or hormones that we are seeking, we have little chance of saying which is the correct one.

15 · 3 : 1 Vernalisation

Mention has been made above of the characteristic need of certain long-day biennials for exposure to low temperatures before they can be induced to flower. In the grasses it is sometimes possible to recognise a ' juvenile ' phase in which the plant is insensitive to the environmental conditions which at a later stage might promote flowering. There follows a second inductive phase in which the plant is, as it were, prepared for the subsequent third phase, viz. the photoperiodic initiation of flowering. However, so far as most *cereal* grasses are concerned the juvenile phase is absent, and the young plant can be influenced with respect to its future flowering from germination onwards. Agronomists are interested primarily in the yield of cereals, especially as it is related to the length of the available growing season. In northern temperate regions many wheats, for example, are winter varieties, sown in the autumn and with a good measure of cold resistance; if they get away to a good start in the spring they may yield well. Spring varieties, on the other hand, generally have a lower frost resistance, but may achieve anthesis earlier, and can be harvested sooner. Varieties may be distinguished according to whether they have a real demand for a cold period (*obligate*), or whether they can manage to flower even if they do not experience one (*facultative*). It is known that winter wheats can be treated so as to make them behave more like spring wheats; the treatment, known as vernalisation, consists of partially moistening the grain and then keeping it for a sufficiently long period (perhaps a month or more) at a suitably low temperature (about 5°C). The grain is sown in the spring, and, like spring wheat, crops sooner than if it had been sown in autumn in the ordinary way.

Plant breeders are better able nowadays to introduce desirable physiological traits into new strains, and they have produced so-called 'alternative' varieties, with moderate frost hardiness and a fairly short season, useful for either spring or winter planting. The farmer is thus enabled to time his sowing according to his local conditions, and still be assured of a reasonable yield. In fact the importance of vernalisation is probably best realised in countries such as the U.S.S.R., where it enables the limits of a given crop to be pushed further northwards into areas where the winters are more severe, and where the growing seasons are shorter.

Seeds are not the only organs which can be vernalised in this way, and of course in the case of the long-day biennials already discussed, the cold temperature treatment is applied to a plant that has already made one season's growth. In general, cold-temperature treatment is only found to be effective where the organ shows the presence of actively meristematic tissue. A generalisation may be stated, viz. that all plants (other than chrysanthemums) that have a chilling requirement subsequently require long-day treatment in order that they may flower.

15 · 4 *Regulatory hormones in flowering*

In discussing the onset of flowering we have, of course, already discussed the production of a stimulatory substance in the leaves, which is transmitted to the apex and there brings about an increase in the number of primordia and a great deal of change in the manner in which these primordia differentiate (pp. 154 and 201).

From this point onwards the inflorescence and its developing flower buds start also to dominate the nutritional scene. They become a major receiving ' sink ' for the consumption of substances translocated from the leaves and elsewhere in the plant. High rates of protein synthesis occur here, and at least in many herbaceous plants, root growth is considerably diminished (if not stopped altogether) whilst so much demand is being made on the capital available.

In addition to this change in the pattern of nutrition, and onwards throughout the whole of flower and fruit development, there is evidence of closely and subtly regulated sequences of processes. The development of the protective and attractive floral envelope still owes something to the more usual vegetative sequences of leaf production and development, but may show new and distinct changes in the pattern of growth. The petals, and sometimes even the sepals, show biochemical divergence from ordinary leaves in the production of sap soluble pigments (e.g. anthocyanins) and chromoplasts, containing carotenoid pigments. The petals show a very much diminished production of chlorophyll. The succeeding sporogenous members (anthers and ovules) are grossly modified in form from the leafy members of the floral envelope, and at present we can only guess about the changes which build up from mitosis to the climax of meiosis, as a result of which *microspores* (pollen grains) and *megaspores* (embryo-sacs) are formed. Fertilisation following pollination, brings with it much more tangible evidence of hormonal activity. Some of the earliest observations on auxin effects were made using aqueous extracts from the pollen masses of orchids. True pollination (or, as an alternative, the deliberate application to the stigma of a synthetic hormone) can induce the setting of fruit and its subsequent development from the young ovary in a plant such as tomato. This implies a control of abscission ; otherwise the ovary would wither and drop, because it had been isolated from its sources of supply by the growth of an abscission layer. The effects of applying synthetic auxins also imply a promotion of all the other aspects of fruit growth besides the development of the ovules. In fact the fruits that develop after auxin treatment are seedless, and are said to develop parthenocarpically.

The process of pollination clearly involves an initial hormonal stimulation by the pollen grains, probably continued as the grains germinate and as the pollen tubes grow down towards the ovules. Certainly the fertilised and developing ovules are a powerful and continuing source of hormone production.

Evidence for this is available from a number of sources but we shall illustrate with only two. Apples may sometimes be seen to be very asymmetrical. If one of them is cut open along a horizontal plane through the ovary, it may prove that the ovules in one or more chambers of the ovary have aborted. This gives some clue as to the way in which hormones are diffusing from the actively growing ovules, and exercising an effect on the growth of the fleshy part (receptacle) outside the ovary walls. Tissues next to the aborted ovules fail to expand and the fruit becomes lopsided. Again, in the strawberry, the degree of development of the fleshy receptacle can be shown to depend on numbers of achenes present (separate one-seeded carpels borne at the surface of the receptacle). One-sided removal of the developing achenes markedly depresses the growth of the receptacle on that side (Fig. 15 · 6).

15 · 5 *Dormancy and the seed habit*

The maturing ovary protects the ovules which themselves meanwhile are maturing into seeds containing the developing new embryos. A good question at this point would be: ' Why do not the embryos in young ovules continue to grow straight on into young plants? ' In fact very few species do so; a classic exception is found in the viviparous mangroves (*Rhizophora* spp.). Here, the embryo extends and gets heavier and heavier, bursting through its coverings and finally dropping from the tree like a plummet, radicle downwards, into the mud of the swamp below; here it continues to grow with little interruption, and there has been no notable pause in the growth process.

However, most embryos cease their intensive early phase of growth activity as the seed matures and becomes dormant. By this time, reserves have been laid down in embryo or endosperm, the testa often becomes more or less hard and sometimes even water-proof, and meantime water has been lost or withdrawn from the seed. In the condition in which most seeds are shed, the tissues are at rest, the respiration rate is very low indeed and the water content has dropped to a minimum consistent with the retention of life. In this

Fig 15·6 The results of an experiment by Nitsch in which the development of a fleshy receptacle by *Fragaria sp.* (strawberry) is shown to be controlled by the developing achenes present, as discussed in the text.

condition the embryo is often quite resistant to adverse conditions. It is claimed, for example, that one can sometimes germinate the ' pips ' from raspberry jam. Certainly seeds show great resistance to extremes of cold and drought. However, if the drying-out process goes too far, which generally means that the seed has been kept from germinating for too long, the seed proteins become denatured and death ensues. Commercial seedsmen are required by law to sell seeds of not less than a certain prescribed viability, and routine germination percentage tests must be carried out on samples in order to ascertain their viability, and to ensure that the current season's yield is not overdiluted with more aged seeds of lower viability. If the seed drops to the ground and becomes washed down a crack, or is otherwise covered with soil, it can absorb water, but for germination to proceed it still requires adequate warmth and oxygen. Thus many seeds are prevented from germinating for lack of water, or by low temperatures in winter, or by waterlogging of the soil, which results in low oxygen levels. This is known as *imposed dormancy*. Given, however, that all the appropriate conditions are available, seeds may often still fail to germinate and may stay in a state of *innate dormancy*, controlled by internal rather than external factors.

There are a number of reasons why an embryo might be held in an arrested state of growth. The factors involved may include those to do with the seed coat or testa, and those concerned with the embryo itself.

SEED COAT FACTORS are of several kinds. At this point we shall deal only with mechanical factors, and factors which control the permeability of the testa to water and gases. At the end of the section will be found reference to chemical substances inhibiting growth (and these are sometimes located in the seed coat) and to light effects upon germination (these again can often be located in the testa).

Some seed coats are so hard and tough that they offer mechanical resistance to the expansion of the embryo and its imbibition of water. The standard example often quoted is *Alisma* (the water plantain) whose seeds may lie, imbibed to their limit, in the mud until bacterial action weakens the testa and allows the embryo to expand without restraint; but it now seems that *Alisma* seeds also need exposure to cold conditions as well. In many of the Leguminosae, the seed is hard and waterproof, and gardeners have adopted the technique, for example in the case of *Lathyrus odoratus* (sweet pea), of 'scarifying' the testa by mechanical abrasion, so that water may enter more freely. (An alternative technique is to soak the seeds for a short time in concentrated sulphuric acid.) A testa that is impermeable to water may sometimes be impermeable also to oxygen from outside, and may hinder the escape of CO_2 from within. In these cases, alternate wetting and drying of the seed may bring about the formation of sufficient micro-fissures in the testa to make the passage of such small gas molecules possible. In some of the clovers (*Trifolium* spp.) the micropyle is hygroscopic and responds to changing humidity in the soil atmosphere by opening and closing;

thus the seeds cannot readily take up water until the right humidity brings about the right degree of opening.

EMBRYO FACTORS are similarly varied, and are often linked to seed coat factors. We have already discussed the very immature orchid embryo (p. 196); *Ficaria verna* (the lesser celandine) provides yet another example of an undifferentiated embryo that needs time to mature morphologically as well as physiologically, before germination can take place. Embryo growth is most frequently arrested by biochemical means, and here may be quoted the classic example of *Sinapis alba* (white mustard), in which a testa factor is also concerned. In seeds of this plant, dormancy is imposed on the embryo by supranormal concentrations of CO_2, and growth is impeded unless the testa is removed, or is mechanically disturbed by alternately wetting and drying the seed.

However, chief of all the causes of embryo dormancy is the presence of growth inhibiting substances; sometimes these are found in the seed coat or ovary wall but often they accumulate in the embryo tissue as it passes into dormancy. In the family Rosaceae especially, gardeners are well aware of the necessity for a period of so-called *after-ripening*; for long they have treated the seeds of rose, hawthorn, apple etc. by an empirical chilling process. Often seeds are put down in layers in sand and left open to the lowered temperatures of winter, though it is a good deal easier to control this treatment using a domestic refrigerator for the chilling process. If held fully imbibed at a temperature of just under 5°C for a suitable period, their metabolism may be advanced to the point at which germination can occur. Many seeds, e.g. *Betula* (birch), *Rumex* and *Polygonum spp.* (docks), respond similarly to an experience of low temperature, and it is convenient to think of these as mature only in a morphological sense, for they are not yet physiologically ' ripe-to-germinate '. It is known that inhibitory substances (e.g. in dormant apple seeds) diminish during chilling; and in a number of other instances it has been shown that the disappearance of inhibitors may be hastened by the application of gibberellic acid. Gibberellin treatments (see p. 166) appear to mobilise enzymes which make hexoses and, more particularly, amino-acids available to the pool of metabolites upon which growth depends. It is also a generally observed fact that gibberellin levels increase in seeds as they emerge from dormancy, and there follows a great stimulation of metabolic turnover.

Although this is a chapter primarily devoted to aspects of the physiology of reproduction, it should be added that some of the features of dormancy, discussed above for seeds, may be found also in vegetative structures (e.g. buds, tubers, rhizomes etc.) that may become dormant as part of the normal seasonal behaviour of the plant. The onset of such dormancy is usually part of a photoperiodic response, as a result of which growth-inhibiting substances build up in concentration at the growing point, and generally with the onset of short-day conditions. With organs such as the buds of woody plants, the tubers of *Helianthus tuberosus* (Jerusalem artichoke) etc. a cold temperature exposure will help to overcome the inhibition to growth. Note, however, that the reverse sometimes holds for dormant potato tubers, which may better be stimulated into growth by exposure to higher temperatures, of the order of 30°C.

Furthermore, it is now possible in a number of instances to infer that the inhibitor is a substance known as abscisic acid; the application of very small amounts of this substance to the leaves of seedlings of woody plants, like *Betula pubescens* (birch), causes them to form resting buds. Release of dormancy results in an increase in the levels of endogenously formed gibberellins, and it is interesting that it is often possible to overcome bud dormancy by the exogenous application of gibberellic acid. In dormant seeds with a cold temperature requirement for germination, there seems to be a similar interaction between an inhibitor, shown in a limited number of instances to be abscisic acid, and endogenous gibberellins. Inhibiting substances are sometimes found in the seed coat or in the ovary wall, and these would normally have to be leached out by rain before germination could proceed. Inhibitors, including coumarin and various phenolic substances, are of common occurrence, and are of special relevance in the desert situation where a very short rain season is preceded by light rains. The light rains are instrumental in gradually washing out the inhibitors, and germination can proceed by the time that the real (though still rather limited) rainfall has arrived. This clearly has survival value.

A relationship between the break of dormancy and light has long been recognised in a number of plants. Seeds that respond to light are known as photoblastic; they are *positively photoblastic* if exposure to light stimulates their germination, and *negatively photoblastic* if exposure to light maintains their dormancy, i.e. if they germinate best in the dark. Examples of positively photoblastic seeds are *Lactuca sativa* (lettuce), the variety Grand Rapids being more sensitive than most, *Epilobium hirsutum* (the large willow herb), and various species of *Rumex* (dock) to name but a few. Negatively photoblastic seeds are less common, but the garden plant *Nigella damascena* (Love-in-a-mist) and the weed *Lamium amplexicaule* (henbit) may be cited as fairly easily obtained examples that could very suitably be used for experiment and trial.

An early discovery by Borthwick and his team implicated the light control of germination with the operation of the phytochrome mechanism, and a great deal of their subsequent findings arose out of the work with lettuce seeds. It is not always clear as to whereabouts in the seed this particular light reaction takes place. In general it might be thought that the tissues of the endosperm or embryo would be the seat of a red/far-red reaction, as a result of which perhaps an inhibitory substance may be changed or removed. Thus the light requirement of lettuce for germination may be replaced by chemical treatments (e.g. with thiourea, gibberellic acid etc.). However, sometimes the isolated embryo will germinate quite satisfactorily without a light requirement, and this suggests that the seed covering is the locus of the operation of the phytochrome mechanism. A great deal of work remains to be done on individual seeds in order to find out whether the light-controlled inhibitory effect located in seed coverings operates through (*a*) changes in the mechanical resistance to embryo expansion, (*b*) changes in the diffusion resistance to the respiratory gases (CO_2 and oxygen), or (*c*) changes in chemical inhibitors located in the seed covering and capable of affecting embryo growth, or combination of any or all of these factors.

15 · 6 *Recapitulation*

Figure 15 · 7 shows the life cycle of a flowering plant in broad outline, and brings together some of the relationships that have so far been demonstrated in the activities of the four naturally occurring groups of endogenous growth regulators during the vegetative and reproductive phases of growth.

It is not yet possible to present a complete picture of the reactions and interactions of these growth regulators in plants, and indeed it may never be possible to do so. Even after the necessary simplifi-

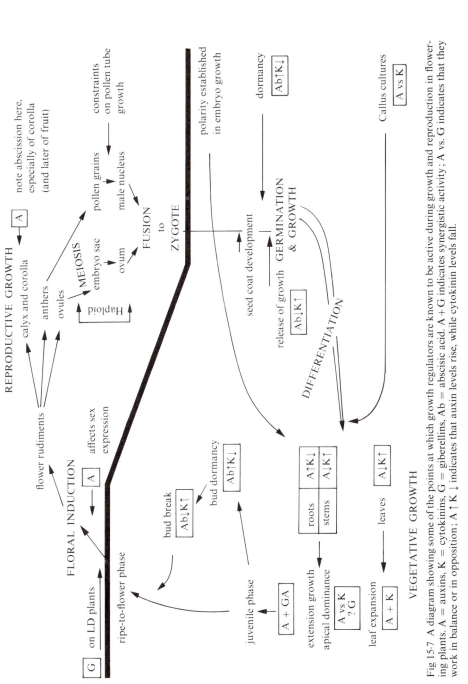

Fig 15·7 A diagram showing some of the points at which growth regulators are known to be active during growth and reproduction in flowering plants. A = auxins, K = cytokinins, G = giberellins, Ab = abscisic acid. A + G indicates synergistic activity; A vs G indicates that they work in balance or in opposition; A↑ K ↓ indicates that auxin levels rise, while cytokinin levels fall.

cation has been attempted, it remains evident that the whole situation is most intricate. It seems unlikely, furthermore, that the situation can begin to be resolved until we have a much clearer idea of the specific mode of action of all of these regulator molecules, whether this proves to be at the enzyme-substrate level, the ribosome-enzyme level, or by direct action at the level of the gene. Indeed there is no evident reason why such an ubiquitous molecule as IAA should not be acting in all three ways.

PART FIVE

Response and survival

16   Adaptation to environment

This book has attempted to review some aspects of the diversity and control of function amongst higher plants and this has been related wherever possible to structure. It is not difficult to present structural and functional diversity in simple terms as offering evidence of the long-term processes of adaptation by plants to an environment that is always changing, however slowly. The new environment may have become more harsh or more beneficial, but some adaptation has always taken place. Thus plants of the coal measures lived in a warm and steamy environment, and their lush growth and frequently large stature can be contrasted with the diminished habits of the alpine flora that is so much more representative of plants that have relatively recently lived in the threat of extinction by cold.

16 · 1   *Morphological and metabolic adaptation*

Much of the anatomical variation recorded here and elsewhere can be regarded as involving adaptation by the plant in relation to its requirements for adequate supplies of water, and this is often coupled with the urgent necessity for water conservation. Other features, such as variation in the epidermis and in the internal airspace system of the plant, can be related to the need for controlled gas exchange, without which photosynthesis in terrestrial plants could never have developed as the efficient process that it clearly is. Furthermore, adequate oxygen supplies might otherwise never be as easily available for energy production, and growth in vital but distant regions such as the root system, were it not for an airspace system which allows for sufficient gaseous diffusion.

Then again, size, whatever the drawbacks at the upper end of the size scale, undoubtedly confers capital advantages on the plant; a greater photosynthetic leaf area can be maintained and thus more reserves can be stored. Furthermore, in larger perennial plants such as trees, heavy reproduction rarely puts a fatal strain on the capital available. This is clearly not the case with most herbaceous annuals; they survive primarily because they have adopted the seed habit as a means of by-passing adverse conditions. They have incorporated amongst their life-processes the metabolic trick of seed dormancy, just as other plants have incorporated the adaptations of frost-hardiness, or of resistance to desiccation.

Thus both morphological and metabolic adaptations in some variety have been indispensable for survival in the face of adverse climatic changes. Yet it must be pointed out that climatic changes do not provide the only source of adversity; for instance, other competing organisms may all too frequently form a living part of the hostile environment. Success and survival often depend on vegetative as well as reproductive vigour, but not least upon ' winning the competition ' both above and below ground level.

16 · 2 *Reproduction*

Quite deliberately, a good deal of time has been spent in Part 4 in discussing the reproductive process, and it can hardly be too often repeated that for any population of a given species, meiosis during sexual reproduction is the key process; it keeps the pool of genes in flux, and allows new and beneficial mutations to be incorporated into the genotype. This clearly offers the best change of adaptation and survival.

Reproduction, as we have seen, may also be linked through dispersal to survival. Thus the effectiveness of the dispersal of seeds must be seen in terms not only of the capital that they carry and the distance that they travel, but also of their capacity to weather adverse conditions and to germinate when their environment at last becomes favourable. Some balance must be struck between the need for economy in the provision of capital reserves in seeds and the advantages of wasteful yet wide dispersal. Is it better, we might ask, to produce fewer and larger seeds, so that, whilst they may travel less far from competition with their parents, they may nevertheless stand a good chance of producing viable offspring with a better chance in life? Or, as in the case of the orchids, is it better to provide countless tiny seeds with the certainty that the bulk of them will perish, though a few will encounter just the right conditions for growth and survival?

16 · 3 *Perception of environmental change*

It would indeed be difficult to study the sequences of flowering, seed set, dormancy and germination without recognising a whole range of processes whereby the plant senses its changing environment and reacts to what it finds. Sometimes the reaction is initiated by a light-sensing mechanism, as when a long-day plant perceives the change in daylength and responds by entering a reproductive phase as the days lengthen. Again, the dormancy that is imposed upon an embryo often results from a daylength response. Sometimes reaction to change may take the form of perceiving an increase in temperature; the response is mediated through biochemical pathways that are more appropriate to the new set of temperature conditions and, chemically speaking, are more probable.

However, in what has gone before, nothing has been said about the recent work that has been carried out on internal metabolic rhythms in plants, and yet knowledge of these seems to offer an important potential weapon in the armoury of those who want to understand how the plant manages to regulate itself within a gradually changing as opposed to a fluctuating external environment.

It has been shown, for example, that detached leaves of the succulent plant *Bryophyllum* have a periodicity of CO_2 emission, with peaks that occur at intervals of about $22\frac{1}{2}$ hours at 25°C. The fluctuations in CO_2 output appear to result from periodic changes in the intensity of fixation of CO_2 in the dark; this feature of *dark fixation* of CO_2 is outstanding in members of the Crassulaceae, the family to which *Bryophyllum* belongs. It also occurs, but to a less well marked extent, in a number of other families. An *endogenous* rhythm of this kind seems to reflect a rhythmically changing metabolic state of the cells and tissues involved, and is best detected by placing the tissues concerned into a 'steady-state' environment, so that fluctuations due to external change may be minimised. Since the rhythmic *period* (or time from peak to peak) is only about a day in length (Latin: *circa diem*) the rhythm is described as *circadian*. Amongst the other examples of circadian rhythm which have been investigated are the leaf movements seen in *Phaseolus multiflorus* (runner bean), with a period of about 28 hours at 25°C. Rhythms like these may be discerned in a number of metabolic processes in addition to respiratory CO_2 production. A number of green algae have been shown to display periodicity in their potential for photosynthesis. One particular marine flagellate, *Gonyaulax*, often responsible for 'phosphorescence' of the sea, displays periodicity in its capacity to luminesce, as well as in its photo-synthetic potential. It seems as if some organisms have their own built-in metabolic timing systems, against which they can detect the effects upon themselves of external cyclic changes (in light, temperature, etc.). Differences may then act as a source of feedback, inducing internal change, and thus as a means of adaptation to the new circumstances. Experimental work in this field has been concerned with the effects of light and temperature treatments on the periodicity of plants (and animals), and more recently with attempts to analyse endogenous rhythms in terms of biochemical change. So far there is a little evidence that links periodicity with quantitative changes in nucleic acids, though not especially with change in any of the specific enzymes investigated; however, very much more remains to be accomplished in this area of research.

The potential for short term as well as long term adaptation in the plant thus depends on its capacity for sensing the environment and responding to change. So that the plant becomes the joint expression of the genotype that it inherits and the environment to which at all times it must be adapting itself. Its response may then be either one of direct metabolic variation with the varying environ-ment (as when the production of energy varies with temperature), or it may be one of diverted metabolism (as when a lowered tempera-ture operates an alternate set of metabolic pathways) (see for example p. 207). Sometimes, but not always, metabolic change arises from the disclosure and operation of previously masked sets of genes. Circadian rhythms offer us yet one more mechanism whereby external change (say in the thermoperiod or the photo-period) can be perceived by a kind of internal matching with the plant's own endogenous rhythm.

The working plant must adapt to survive, and, no less than its more mobile and more complicated animal counterpart, it requires the means of interpreting its environment. Without these means it can truly do little more than respond physically to the forces at work upon it.

Selected books for reference

Textbooks

CLOWES, F. A. L. (1961) *Apical meristems*. Blackwell Scientific Publications, Oxford

CLOWES, F. A. L. and JUNIPER, B. E. (1968) *Plant cells* Blackwell Scientific Publications, Oxford

ESAU, K. (1961) *Anatomy of seed plants* J. Wiley & Sons, New York & London

KOZLOWSKI, T. T. (1964) *Water metabolism in plants* Harper & Row, New York

HILLMAN, W. S. (1962) *The physiology of flowering* Holt, Rinehart & Winston, New York & London

LEOPOLD, A. C. (1964) *Plant growth and development* McGraw-Hill, New York

SINNOTT, E. W. (1960) *Plant morphogenesis* McGraw-Hill, New York

STEWARD, F. C. (Ed.) (1959 et seq.) *Plant physiology* Academic Press, New York & London

SUTCLIFFE, J. F. (1962) *Mineral salts absorption in plants* Pergamon Press Ltd., Oxford

WAREING, P. E. and PHILLIPS, I. D. J (1970) *The control of growth and differentiation in plants* Pergamon Press Ltd., Oxford

WILKINS, M. B. (Ed.) 1969 *Physiology of plant growth and development* McGraw-Hill, New York & London

Shorter references

BURGES, A. (1958) *Microorganisms in the soil* Chapter 7 The role of micro-organisms in the cyclic systems Hutchinson & Co. Ltd, London

RUSSELL, R. S. (1970) ' Root systems and plant nutrition; some new approaches ' *Endeavour* **19**:84

STEWARD, F. C. (1963) ' The control of growth in plant cells ' *Scientific American* **209**:104

STEWARD, F. C. (1970) ' Totipotency, variation and clonal development of cultured cells.' *Endeavour* **19**:117

Index

Numbers in italics refer to figures.